Praise for *Aligned*

"The one most important skill to be successful is the one they don't teach you in university: how to collaborate with other people. This book is the course you wish you could have taken; Bruce and Melissa cover negotiation, collaboration, persuasion, and much more. If you struggle with getting buy-in, can't seem to get things done, and agree with Sartre, 'Hell is other people,' then this book is your life raft. It's going straight from the printer and into my Stanford syllabus."

—*Christina Wodtke, author of* Radical Focus

"Most of your work as a product leader is aligning people (and keeping them aligned). *Aligned* is the best book I've come across to help build this critical skill."

—*Lenny Rachitsky, author of* Lenny's Newsletter

"*Aligned* gave me a touch of PTSD! The characters and situations brought me back to challenges I faced throughout my career. I wish I had this book to provide context, understanding, and useful frameworks to navigate a successful path. *Aligned* is essential reading if you happen to work with people."

—*Jason Moens, former VP Product, Flywire and Toast*

"Product leaders operate in the realm of ambiguity, whereas the rest of our organizations crave certainty. This book provides essential tools for building effective stakeholder relationships across that divide."

—*Matt Kaplan, former SVP Product, Toast*

"The ability to drive alignment across the organization is critical for a product executive—really any executive—to ultimately drive growth. This book distills the authors' years of hard-won wisdom into actionable tips and wraps it in a compelling story that will resonate with anyone who has ever tried to lead a team."

—*Irena Goldenberg, cofounder and partner at Highland Europe*

"Bruce and Melissa have written the missing manual for product leadership. It is mandatory reading for any product person."

—Anabela Cesário, EVP of Product Management, Outsystems

"A strategy is only as good as the team behind it. This book provides the hands-on tools and techniques every product person needs to make their visions reality."

—Marcus Bittrich, CPO, NewStore

"Regardless of industry or job title, alignment with people who have a vested interest in your product is key to success. Thank you, Bruce and Melissa, for sharing your wisdom and experience, and providing product professionals a much-needed resource on this topic."

—Jerry D. Odenwelder Jr., Product Coach, Cprime

"Every product manager at every level has challenges in gaining alignment with their stakeholders. *Aligned* offers PMs hope and an approach that they can work through to build a coalition and get everyone in their organization moving collectively in the right direction. I wish I had this book at the beginning of my career!"

—C. Todd Lombardo, Product VP and coauthor of Product Roadmaps Relaunched

"Product managers live and die by their stakeholders, from team members to executives. *Aligned* is the reference for product people on stakeholder management, an invaluable guide to help you get even the most difficult, stubborn people on your side. Don't go without it. An invaluable guide for product people with even the most challenging stakeholders."

—Tim Bouhour, Product VP

"Alignment is the missing piece of the 'how to create great products people love' puzzle. If you are seeking product seniority, it's time to face the politics and learn to wrangle stakeholders effectively, and you finally have an instruction manual. *Aligned* doesn't just talk about the problem—it's the most practical book I've read in years."

—Tamara Adlin, Early Stage Advisor and President, Adlin, Inc.

"*Aligned* is an invaluable guide for new and seasoned product managers alike seeking to master the art of stakeholder management. With a fresh, pragmatic approach and practical strategies, real-world examples, and a focus on building trust and alignment, *Aligned* delivers a comprehensive and fresh approach. It's a must-read for anyone looking to excel in their product management career."

—Michael Pierce, Director of Product Management

"Bruce and Melissa expertly crafted the perfect blend of concise and practical guidance, complemented by a profound comprehension of the underlying principles. Their guidance possesses an undeniable authenticity, enriched by a mix of personal anecdotes from both themselves and their peers. Moreover, they have devised effective exercises that can help even the most seasoned product managers."

—Liz Lehtonen, Head of Production, NetEase First-Party AAA Studio

"*Aligned* is a book that I wish I had earlier in my career, as it would have helped all those 'good ideas' get a meaningful chance at assessment, instead of being a lesson in how to talk to stakeholders. And sometimes those lessons were painful. If you want your influence to work outside of your closest friends, *Aligned* is the book you need to pick up today."

—Adam Thomas, Principal, Approaching One

Aligned

Stakeholder Management for Product Leaders

Bruce McCarthy and Melissa Appel

Beijing • Boston • Farnham • Sebastopol • Tokyo

Aligned

by Bruce McCarthy and Melissa Appel

Published by O'Reilly Media, Inc., 1005 Gravenstein Highway North, Sebastopol, CA 95472.

O'Reilly Media books may be purchased for educational, business, or sales promotional use. Online editions are also available for most titles (*https://oreilly.com*). For more information, contact our corporate/institutional sales department: (800) 998-9938 or *corporate@oreilly.com*.

Acquisitions Editor: Amanda Quinn
Development Editor: Angela Rufino
Cover Designer: Michael Connors
Mechanical Designer: Susan Brown
Interior Designer: Michael Connors
Production Editor: Kristen Brown
Copyeditor: Liz Wheeler
Proofreader: James Fraleigh
Indexer: Potomac Indexing, LLC

June 2024: First Edition

Revision History for the First Edition

2024-05-31: First Release

See *http://oreilly.com/catalog/errata.csp?isbn=9781098134426* for release details.

ISBN: 978-1-098-13442-6
[LSI]

Contents

It's about leading your team to achieve something great together

Preface

In the dimly lit pre-dotcom days of 1996, Bruce developed a fabulous new internet-based product for marketers. Before print on demand, before Mailchimp, before HubSpot, and before anybody really had a good sense of how to make money on the web, there was his product: LetterBuilder.

It was simple. It elegantly solved a key customer problem. It was easy to buy and to use. It got his company a toehold in this new frontier, the world wide web. And it was doomed from the start.

Bruce had worked out the strategy, validated the problem and the solution with customers, worked out the pricing, created a partnership, written the requirements, hired a design firm, and worked with the developers and the testers. His plan was foolproof. He was creating the first virtual print shop for direct marketers years before anyone had heard of VistaPrint, and he was going to single-handedly propel his 50-person startup to dotcom stardom and IPO.

What killed his brilliant product? He forgot about the rest of the company.

Bruce kept everyone in the loop with updates on progress with emails and status reports. His boss, the VP of product, was enthusiastic about the idea. But when Bruce met with the VP of marketing a few weeks before the launch, she asked him what his plan was for promoting the product.

Wait, *Bruce's* plan? Wasn't marketing *her* job? Thinking fast, he pitched some ideas of things her team could do to help get the word out and maybe bundle it with the company's core offering. She was unmoved.

It turned out that the entire marketing budget for the year was already allocated to lead generation and renewal efforts for the company's core money maker. The VP of marketing thought what Bruce was working on was interesting, and acknowledged that it solved a problem for the customer, but was crystal clear that it did absolutely nothing for *her* goals. She could not afford to divert resources from the programs that drove the company's growth quarter after quarter for his fledgling effort, no matter how promising.

That was bad. It got worse.

The compensation plan for sales had already been fixed and the VP of sales was not interested in complicating it with a low-priced add-on when it was easier to sell more units of the core product. But the *customer*, Bruce argued. But our *web strategy*, he pointed out. "Look," the VP of sales said, "I want to help, but I have a number to hit and this isn't helping *me*."

The very next day Bruce's boss and the CEO took him to lunch to celebrate the impending release. In the car on the way to the restaurant, the CEO asked Bruce what his expectations were for the new product. He told him frankly that he had no expectations at all given the responses he'd gotten from sales and marketing.

It was a very quiet lunch.

Why We Wrote This Book

Just like Bruce, many product managers linger inside their silos until it's too late. It's easy to forget that you need to align with many important people in your organization to drive successful products. Even if you know you should talk with them, maybe you're not sure how or when. Or maybe you know how to engage with them, but there's one person in particular who you dread talking with.

Maybe you work with a CEO who thinks she knows what product management is, but asks you to be responsible for getting the engineers to deliver code on time. Or your VP of sales insists you drop everything to deliver a seemingly arbitrary feature because his prospect "won't sign without it." Maybe you're always responding to the latest executive "emergency" and it's hard to catch your breath, let alone come up with next quarter's roadmap. Or maybe, like Bruce, you've got a brilliant plan but no one will listen.

There *is* a better way.

We wrote this book to help you—and your product—be successful. By reading this book, you will learn how to do the following:

- Map your organization and identify Power Players.
- Build rapport and trust with stakeholders.
- Leverage curiosity, preparation, and expectation-setting.
- Create focus to deliver real customer value.
- Minimize roadmap derailment and guesswork.
- Have stakeholders appreciate and support your "no."
- Sustain alignment over time.
- Manage particularly difficult stakeholders.

Who This Book Is For

If you believe that stakeholder management is key to your career success, this book is for you. Better stakeholder management can help you unlock your next career step, enhance your impact at work, and drive more successful outcomes for customers. Specifically, it is useful for people in the following types of positions:

- Anyone who feels they could accomplish more with enhanced cooperation and collaboration among their coworkers
- Anyone who is having trouble working with their stakeholders
- Anyone who has been told they need to build better relationships with stakeholders but is not sure how to get started
- Individual contributors who want to excel in their current role, who are starting a new role, who want to be promoted, or who are trying to get their first management role
- Managers and directors who want to improve their skills and expand their influence in their organization
- Experienced professionals new to leading their function or joining an executive leadership team for the first time
- Senior leaders, VPs, and CPOs looking for a review of stakeholder management techniques and fresh ideas
- Consultants and coaches who advise clients and find stakeholder management a frequent topic of conversation

Regardless of your company size, building alignment is critical to your product's success. Whether you're at a giant conglomerate, or a tiny start-up, there will be times when people disagree. Understanding how to work with stakeholders can unblock your road to success.

This book is designed for people in product management roles, but the concepts are more broadly applicable. Aspiring product managers and anyone who works on a team that creates products will learn a lot too, such as engineers, designers, user researchers, and analysts. While we focus our examples in this book on product managers, the truth is that nearly everyone in every discipline has stakeholders of some kind. We hope that anyone can find this book useful for improving how you interact with your coworkers as you drive toward alignment.

Definitions

Throughout the book we use the following key terms.

Product Manager

Most people who picked up this book probably know what a product manager is, but just in case you don't, here is how Tony Fadell describes the role in his book, *Build* (Harper Business, 2022).

> [The product manager] is a needle in a haystack. An almost impossible combination of structured thinker and visionary leader, with incredible passion but also firm follow-through, who's a vibrant people person but fascinated by technology, an incredible communicator who can work with engineering and think through marketing and not forget the business model, the economics, profitability, PR. They have to be pushy but with a smile, to know when to hold fast and when to let one slide.

Product management has different definitions at different companies, and, although it's been around for some time now, product management as a profession does not yet have a mature, globally agreed-upon definition. As described in Fadell's quotation, the product managers in this book have a nuanced job. They do not simply collect requests from around the organization and feed them to engineering for execution; rather, they are the hub of cross-functional activity that drives successful products. And of course, their job includes stakeholder management.

A product manager may not be the CEO of the product, as some have claimed, but they are often the only other person in the company who is responsible for integrating customer desirability, technical feasibility, and business viability.

Product Team

"Product Team" is a term given to a small cross-functional team that works on a specific product or piece of a product. Sometimes the Product Team is called a "pod" or a "squad" or a "sprint team," but in this book we will call it a Product Team. The Product Team generally consists of a product manager, a designer, an engineering lead, and several more engineers. Some Product Teams contain other functions like user research or product analytics. In this book, we will simplify the team as just product management, engineering, and design, but we recognize that your Product Team may look different.

Stakeholder

A "stakeholder" is anyone who has an interest in the success or failure of your product or whose support is required for that success. There are a few key categories of stakeholders: internal stakeholders (such as your Product Team, executives, sales, marketing, customer support, finance, operations, and legal teams) and external stakeholders (such as customers, partners, and vendors). While all those types of stakeholders are important for the success of your products, this book will focus on your internal stakeholders outside of your Product Team.

Alignment

"Alignment" doesn't mean getting everyone to do what you want. It also doesn't necessarily mean everyone agrees or gets what they want. Alignment does mean that people feel involved in the process, understand the desired outcome, and commit to working together to make the product successful.

Someone is aligned if they see value in what you are doing, actively want you and your product to succeed, and will help you to accomplish your objectives.

Alignment isn't so different from negotiating. All sides need to understand each other's perspectives in order to have a fruitful conversation. Achieving alignment requires listening, taking in diverse perspectives, proposing a plan, and adjusting that plan until alignment is achieved.

Stakeholder Management

We define "stakeholder management" as the process of ensuring alignment among your stakeholders. It is about leading your broader team to achieve something great together. As product managers we often make final decisions ourselves, but if they've done it right, our stakeholders are already fully aligned.

Stakeholder management isn't stakeholder appeasement. We're going to go out on a limb and say that if stakeholders all feel like they "got something," the product is likely to fail. Effective stakeholder management is an iterative feedback process to drive product success. Product managers act as expert orchestrators, ensuring that their stakeholders understand the needs of the customers, the business, the technology, and each other so that decisions can be made and plans can be executed in harmony.

How This Book Is Organized

To illustrate the advice in this book we alternate between explanations of tools and frameworks and a continuous narrative, starring a composite character named Irie.* Irie is a mid-career product professional learning to navigate the complex environment of a new company. In each chapter, Irie faces new challenges, learns along with the reader, and then applies what she's learned to address her challenges.

Understanding Context

Chapter 1—Organization

- Use different organizational structures to foster more effective stakeholder interaction.
- Understand the difference between the reporting org chart and the influence org chart by identifying Power Players and hidden Power Players.
- Use the Power/Alignment Grid to decide how to move different types of stakeholders into alignment.
- Approach stakeholders differently based on the predominant decision-making culture.
- Design an optimal decision-making authority structure using DACI.

Chapter 2—People

- Prepare for and conduct stakeholder discovery interviews.
- Identify and work with different stakeholder decision styles.
- Find time to meet with hard-to-reach stakeholders.

Building Relationships

Chapter 3—Rapport

- Be relatable, connecting with other people at a human level and exploiting a natural human bias we have toward people similar to ourselves.
- Build mutual respect by assuming good intent, accepting people's differences, and acting with curiosity instead of becoming defensive.
- Practice empathy by validating and sharing in someone's emotional experience.
- Encourage vulnerability, not by oversharing, but by talking about what's really going on, being willing to take on risk, and accepting the possibility of failure.

---

* Irie is a common Jamaican woman's name, pronounced "EYE-ree." It is also a popular expression on the island meaning "all is well."

Chapter 4—Trust

- Demonstrate expertise without sounding like a know-it-all.
- Show confidence, being prepared when you speak with your stakeholders and providing transparency into your processes.
- Take ownership of your product, identifying risks early and proposing mitigations, and owning both your successes and your failures.
- Become a dependable resource, being responsive and delivering on your promises, while setting the right expectations up front.

Achieving and Maintaining Alignment

Chapter 5—Roadmap

- Derive product objectives from organizational objectives.
- Use workshops to drive alignment.
- Mine for conflict, uncovering hidden misalignments.
- Develop a product roadmap that focuses on customer needs and product objectives.

Chapter 6—Changes

- Deal with ongoing requests.
- Plan for routine roadmap updates.
- Decide whether to say yes or no to a roadmap update.
- Say "no" tactfully and effectively.

Overcoming Adversity

Chapter 7—Challenges

- Check yourself and the context to verify that your stakeholder is truly difficult.
- Set relationship goals to deal with difficult stakeholders.
- Evaluate your future options.
- Set quitting criteria to decide how to move forward.

Epilogue

- Three years later...

Visit *alignedthebook.com* for tools, templates, and additional reading referenced throughout this book.

Our Approach to Diversity

Throughout this book, we have made the tips, techniques, and frameworks we present as real and actionable as possible with examples drawn from our experience, the experiences of friends and colleagues, and quotations from other practitioners we have interviewed. (See Acknowledgements for more details on these folks!)

Building relationships at work can be tricky, and not all techniques work for all people. When you go from theory to practice, there are many different factors, such as the unique personalities and backgrounds of the people involved. For example, while writing this book, we—Bruce and Melissa—discovered that gender influences how we think about stakeholder management.

Early on, Bruce wrote a draft chapter about getting to know your stakeholders on a personal level, such as meeting them for coffee outside of the office. Melissa read this chapter and felt uncomfortable with the advice. Because she is a woman, she might not want to invite a male coworker to meet her alone outside the office for fear of misunderstanding or the impression of impropriety. Through this example, we solidified our theory that our perspectives represent only a small portion of the experiences that people have in their day-to-day dealings with stakeholders (and other people) at work.

These different experiences are not limited to gender, race, ability, or other outwardly visible identity factors. Other factors, such as the way people think about problems, their communication styles, or their economic, cultural, or educational backgrounds can also impact how easy or difficult it is to build relationships between two specific people. Studies have shown that diverse teams are more effective than homogeneous teams. By understanding and embracing our different strengths and learning how to build working relationships with folks who are different from ourselves, we make the team stronger and improve our decision making.

With this in mind, we have combined our own experience with interviews and examples that involve people of various ethnicities, genders, and backgrounds. To double-check our advice in this book, we have also tried to be as inclusive as possible when asking for folks to review the book before publication. We have included options and variations on our advice where appropriate to account for different people's experiences and approaches, and have done our best to be inclusive of different people's needs. Everyone's situation is unique and you should take from our advice what works best for you.

Irie Manages Stakeholders

Irie's fictional story is used throughout this book to demonstrate key concepts, using relatable characters and examples. Each chapter's subsections start with Irie's story, then explain concepts she can use to solve her problems, and finally return to Irie to see how things turn out. Here is a partial org chart of Helthex, the fictitious company where Irie works, and an overview of the characters she'll interact with during the story.

Cast of Characters

For your reference, these are the characters appearing or referred to in *Aligned*.

Name	Role
Alex	Analyst relations
Arianna	Director of commercial accounts
Christina	Product manager (PM) for analytics
Darius	Irie's former boss, now an advisor to tech companies
Declan	Data scientist, reports to Divya
Divya	Director of data science
Eitan	Product manager (PM)
Ella	Chief revenue officer (CRO)
Irie	Director of product management
José	Director of design
Justin	Finance associate
Liandri	Director of customer support
Liz	Chief executive officer (CEO)
Min	Product manager (PM)
Mo	Lead engineer
Philippe	Chief financial officer (CFO)
Pria	Executive assistant
Sam	Lead engineer
Sergey	Vice president of marketing
Sparks	Strategic partnerships
Sri	Chief technology officer (CTO)
Wei	Lead engineer, works with Eitan
Yacob	Vice president of engineering
Zola	Customer support manager

Helthex Org Chart

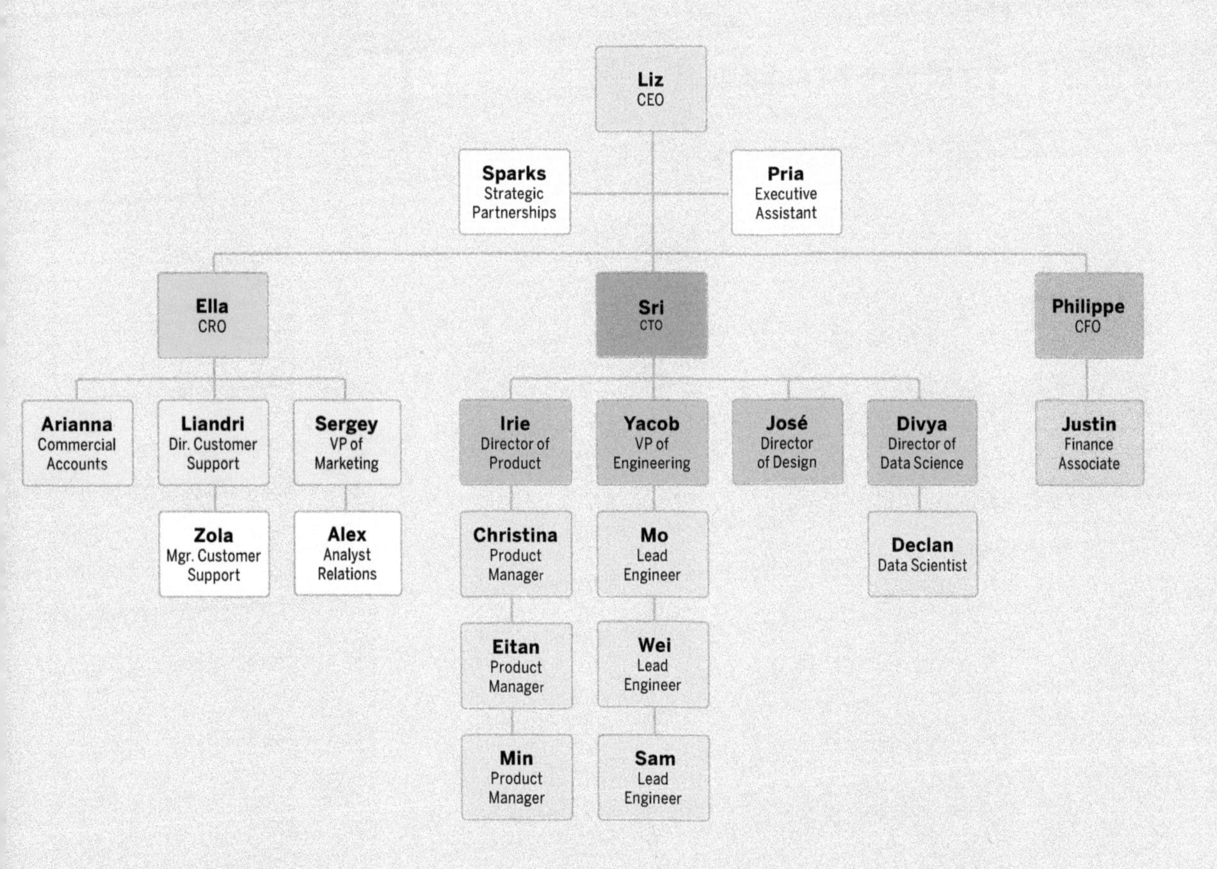

Our Product Management Hero, Irie

Like all of us, Irie is not perfect and she makes mistakes. Throughout the story, though, she learns new techniques and improves her stakeholder management skills. We hope her concrete examples will help you understand how to apply the approaches laid out in this book. We begin Chapter 1 with Irie's first day at her new job…

The reporting org chart isn't the influence org chart

Chapter 1

Organization

Getting to know your organization will help you better identify your stakeholders and will provide context to your interactions with them. How the organization is structured, who influences decisions, and how the decision culture operates all inform your efforts at driving alignment. As our product manager protagonist, Irie,* tries to navigate her new org, we will discuss how to hone some key skills:

- Use different organizational structures to foster more effective stakeholder interaction.
- Understand the difference between the reporting org chart and the influence org chart by identifying Power Players and hidden Power Players.
- Use the Power/Alignment Grid to decide how to move different types of stakeholders into alignment.
- Approach stakeholders differently based on the predominant decision-making culture.
- Design an optimal decision-making authority structure using DACI.

Let's begin by following along as our product manager hero, Irie, learns about the new company she's just joined.

---

* See the Preface for an explanation of how this book is organized (page xx) and an introduction of Irie, our product management hero (page xxiv).

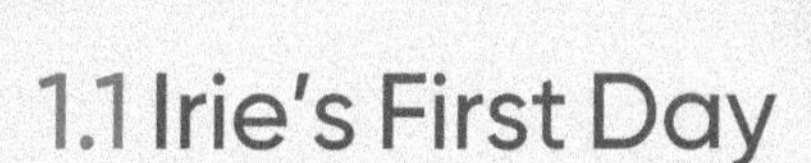

1.1 Irie's First Day

Irie enters the glass-walled conference room, her heart pounding with a mix of excitement and nervousness for the first meeting of her first day at Helthex. As she takes a seat at the table, the room falls silent, and all eyes turn to the CTO, Sri, who sits at the head of the table holding a stack of index cards and displaying a bright, toothy grin.

"Good morning, everyone!" Sri's voice rings out, breaking the silence. Looking at the top card in his stack, he continues, "Today, we have a special introduction to make. Please join me in welcoming Irie, Helthex's new director of product management."

Most of the people around the table—and even the woman on a large monitor at one end of the table—clap and offer warm smiles, making Irie feel slightly more at ease. *Sri didn't have to make a point of my title*, Irie thinks. *But they hired me for my experience.* The three product managers she will now manage each have less than two years' experience.

Given the intimate gathering, Irie decides not to stand, but straightens her posture and addresses the room with what she hopes will be a friendly smile. "Thank you, Sri. Hello, everyone," she begins. "I'm Irie. I'm truly excited to be here as a part of this talented team. I'm looking forward to collaborating with all of you and achieving great things together."

Murmurs of approval ripple around the room, and then Sri asks everyone to introduce themselves to Irie.

Her first day coincides with what Sri calls a "product update meeting" to kick off the quarter. Yacob, the VP of engineering, is there, as are José, director of design, and Divya, leading data science.

The woman on the screen introduces herself as Liandri, saying she manages customer support. Unexpectedly, another voice rings in from the monitor, represented only by a large letter "E." Ella introduces herself as the chief revenue officer, which Irie assumes means she leads both sales and marketing.

The last person in the room introduces himself as "Sparks," explaining that he works on partnerships. Irie is unsure why someone from partnerships is in a product meeting, but it is her first day and she assumes this will become clear in time.

Yacob begins reviewing the work under consideration for their upcoming sprint, opening ticket after ticket in their workflow tool on the meeting room screen for everyone to see. This seems to Irie like a lot of granular detail for a meeting with these high-powered executives. The tickets focus on design enhancements, but there is no discussion of why these changes are needed. Perhaps this group has already agreed on that. At least the tickets are fairly well-formed, Irie observes.

She notices that Liandri and Sparks seem bored, though, and possibly a bit impatient with the discussion. After a few minutes, Liandri breaks in, apologizing. "I have to take this customer call, but can you tell me first if those escalations we prioritized will be addressed this sprint?"

"We got your list," replies Yacob. "We'll try to get to as many as we can."

Liandri doesn't seem satisfied with this answer, but before she can follow up, Ella, the CRO, chimes in. She sounds a bit robotic, like she has a bad mobile connection. "I really need to know if we can commit to the items from the RightBank RFP this quarter."

Yacob explains that they haven't had time to scope that work yet, promising to get back to Ella by the end of the week.

Sparks—the "partnerships" person, as Irie recalls—speaks up for the first time and takes a hard tone. "I don't see the AI assistant we talked about last week in this list. Shouldn't that take precedence over making the UI pretty?"

After an awkward pause at the word "pretty," Sri speaks up. "It's a great idea, of course, Sparks," he begins. "But we don't really know what we want to use AI for. It's hard to know where to start."

With that, Yacob looks pleadingly at Irie, as if hoping she might jump in and save them on her first day. Irie now recalls Yacob and Sri asking her about her experience in handling "politics" in her interviews. *This may be what they were referring to,* she thinks.

Sparks follows Yacob's gaze to Irie briefly, then returns to Sri, sucking in air in preparation for a forceful follow-up.

Irie's enthusiasm wavers, caught off guard by Sparks's challenging tone. She takes a deep breath, steadying herself. Her instincts compel her to try to be helpful and come to the support of engineering and design.

"Sparks," Irie begins before he can press the engineering leaders further, "I'm new here, obviously. Gathering insights from customers and stakeholders—including our partners, of course—is vital to developing an effective product strategy. If the market needs and the precise use cases are not yet clear, perhaps we need to do a bit of homework before we commit the team to this new initiative." She can see everyone looking at her now. She isn't sure, but she thinks she sees approval in a few faces.

"Fortunately," she adds, gaining steam, "I have a lot of experience with customer res—"

Sparks cuts her off, his smirk turning condescending. "Listen, Irene, I appreciate your eagerness to make your mark, but for now, just take my word for it that this is a huge opportunity."

"My name is Irie."

"What?" asks Sparks, confused.

"My name is pronounced 'EYE-ree.' Not 'Irene,'" she says, then adds, "No worries. It's Jamaican, not that common."

"Okay, sure," Sparks says dismissively. "Anyway, it's not really that complicated to integrate an AI. All of our competitors are doing it, or will be soon. We've got to get going on this and we can't afford to waste time working out every detail first."

Irie feels her face get hot with a mixture of frustration and determination. Apparently, she has underestimated Sparks's position at Helthex, and she feels she must salvage a bit of her own credibility. She isn't sure exactly what his role is, but she's dealt with impatient executives before. She feels they often dismiss the hard work that goes into great products in their desire to strike quickly.

"I recognize your concern to move fast, Sparks," Irie replies, her voice steady. "Thanks for that. This is my first day, of course, but the team has clearly worked hard on these design enhancements, and I assume that's for good reason." She looks around the room for support.

José, the design leader, brightens up and offers, "Yes, actually, we're hoping to improve engagement with some of our advanced features by making them more discoverable. You see, people who use these features tend to—"

José is unable to finish his explanation as Sparks interrupts again. "With all due respect, Irie," he says with mock courtesy, "you don't know what you're talking about. It's probably better if you stay in your lane and keep quiet until you understand how things work here."

The room falls into an uncomfortable silence, everyone observing the tense exchange between Irie and Sparks. Sri, recognizing the need to defuse the situation, steps in.

"All right, let's take a step back here," he says, his voice calm but authoritative. "We're all on the same team. Our goal is to deliver outstanding products. Why don't we give Irie a little time to gather all the input and we'll come back to you?"

Ella and Liandri are quiet. Sparks looks like he is tempted to fire back, but now seems conscious that all eyes are on him. "Fine," he offers. "I just want to make clear how important this is to the future of the company."

Sparks leaves the room and the two remote participants hang up, effectively ending the meeting.

Irie is concerned. What has she gotten herself into? She hopes she hasn't made an enemy in her first meeting. Irie is used to the head of product being in charge of product decisions. *But maybe that's not the way things work here*, she thinks. *I need to learn more about how this organization works before I can make a plan for the product. Hopefully, I'll be able to influence Sparks once I learn more about his role.*

Later that afternoon, Irie has her first one-on-one with her boss, Sri, the CTO. They are sitting across from each other at a small four-person conference room table.

Irie notices Sri looks tired and perhaps a little pale. He is sipping some sort of herbal tea. Irie asks if he's feeling okay.

"I just hate big meetings," he says. "I find them exhausting. What do you think of Helthex so far?" he continues, changing the subject.

"It's…a bit different than what I'm used to," says Irie, pausing to find the right words.

"How so?" asks Sri.

Irie begins by revisiting the product update meeting from earlier in the day. "I'm used to different departments advocating for their priorities with the product management team, but it is unclear how these product decisions are made," she says. "Also, I more or less know what to expect from sales and customer support, but I was surprised by how much influence partnerships apparently has. I was at my last company for five years and I guess I forgot that different companies are organized in different ways," she adds.

"Well, let's start with the org chart," says Sri. He brings it up on his laptop and shows it to Irie.

"So what's Sparks's role?" asks Irie, squinting intently at Sri's tiny laptop screen. She finds Sparks up at the top, a single box reporting to the CEO, Liz.

"His role is complicated, and it's changed over the years," says Sri. "I've been here for about six years and I think Sparks has had four different titles. His original role was cofounder."

Figure 1-1. The Helthex org chart

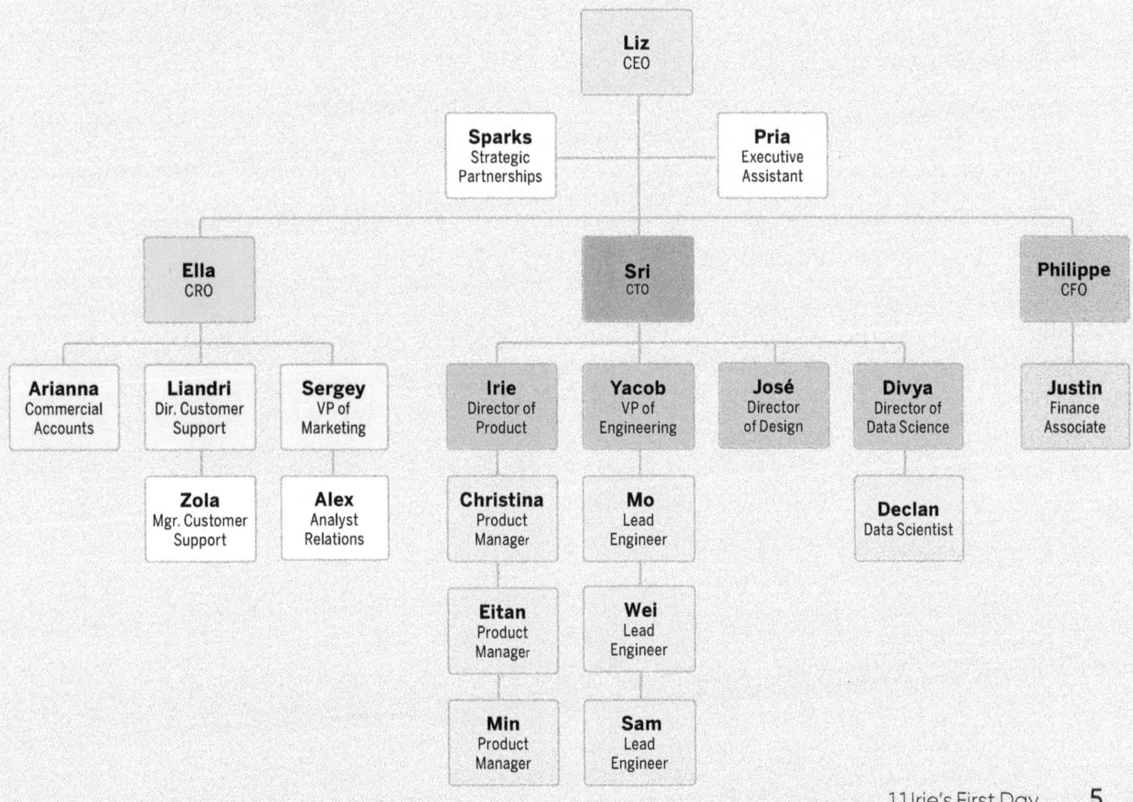

"Really?" asks Irie. "What's his current role?"

"Partnerships." says Sri. "He's always trying to make deals with various partners for a bunch of different things, like packaging our product with another company's offerings or pulling data in from them. Half the time I'm not sure what he's working on; he keeps it a bit hush-hush. But now with Liz out on medical leave, he's taken a lot more interest in the product."

"Yeah, I noticed," says Irie with a smile. "I'd like to get a sense for how decisions are made here, especially product decisions, but can I ask a few more questions about the org chart first?"

"Fire away," says Sri.

"The org chart doesn't show it, but are there any cross-functional teams?"

"Well," says Sri, "our Product Teams are each cross-functional. All three have a product manager, a designer, and three or more engineers. But I don't see that anywhere else in the org. Then again, I wouldn't really know," he adds. "Why is it important how the organization is structured?"

Irie is struck by Sri's lack of knowledge here, but perhaps he tends to keep his focus on the functions he leads—engineering, product, design, and data science. There doesn't seem to be a huge amount of interaction with the other functions, like sales and marketing, outside of the product review meeting. This is her first clue that Helthex might be a "Functional" organization.

"How the different functions interact," begins Irie, "can affect how decisions are made and which stakeholders have the most influence. This drives who makes decisions about the product, which clearly we both have a vested interest in."

"Interesting," says Sri. "I'd love to learn more."

Irie opens her laptop and brings up a slide deck with a slide called "Organizational Structure." She turns the laptop around so Sri can see it.

Sri starts scrolling through the slides. "What is this, like a secret product management guidebook?"

Irie, now somewhat embarrassed, says, "Yeah, basically. Some product management friends and I put this together as a playbook when we were all working together years ago. We've all been adding to it over time, and it's just sort of grown."

"That sounds really helpful," says Sri.

"It *is* really helpful," says Irie. "It doesn't have everything in it yet, because we're always learning new things. But a former colleague joined a really big, complicated organization with lots of moving parts, then added this after she figured it out."

Irie goes on to explain the impact of different org structures on the product. ■

How Org Structure Impacts Product Teams

If you, like Irie, are having trouble navigating your organization, you are not alone. Titles are vague, reporting lines seem to cross each other and double back, and individuals can report to one person, be on someone else's "team," and be on loan to a special project all at the same time. Loyalties can be opaque, and people often claim all sorts of authority that they may or may not actually have.

By understanding how the organization is structured, you can better understand which are the most critical stakeholders to build relationships with and where loyalties lie. Understanding your organization's structure is also fundamental to understanding who has the real power in the organization, how information is shared, and who makes the big decisions. Figure 1-2 shows the four most common organizational structures and Figure 1-3 provides more detail for each one.

Figure 1-2. Most Common Organizational Structures Overview

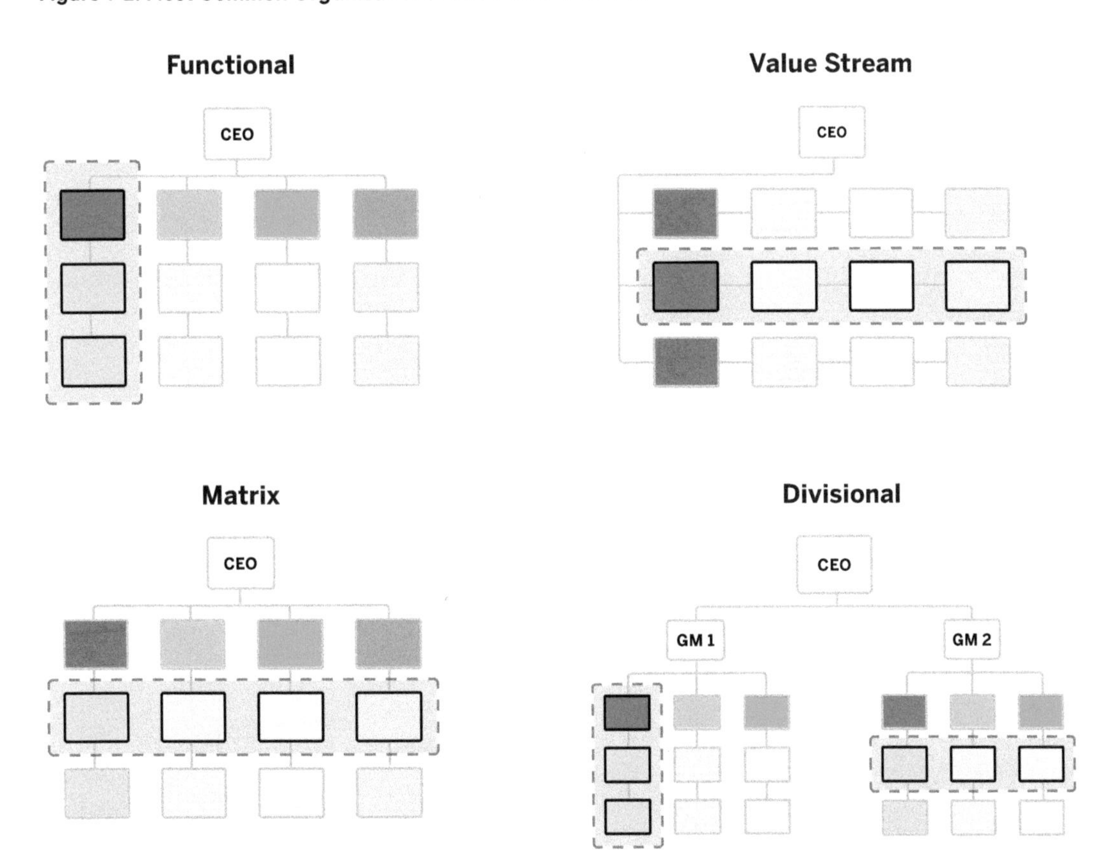

Figure 1-3. Most Common Organizational Structures

Functional

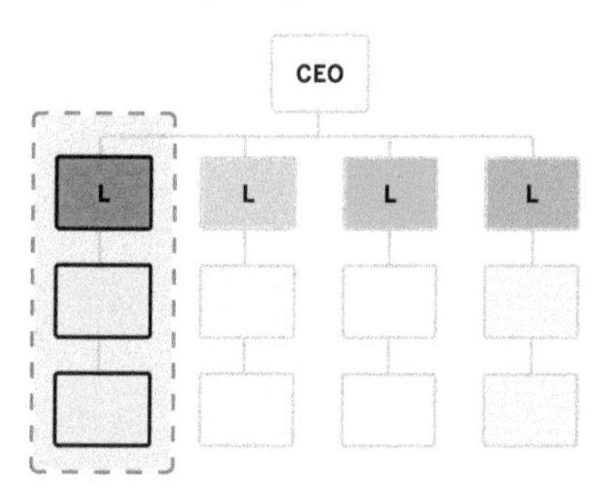

Examples

Starbucks is organized into functional departments such as HR, finance, and marketing.

Apple's technology functions, like design, hardware engineering, and software are separated in the same way as their business functions, like marketing and finance.

Key Stakeholders

Leaders of individual functions such as Marketing, Finance, and HR

If you ask an engineer for a feature

"Let me check with my manager."

Key Metrics

Focus on efficiency, effectiveness, and scale within a function, e.g., cost per lead in marketing.

Benefits

- Functional specialization drives efficiency and maximum compatibility among products
- Clear objectives and accountability
- Leaders coach in their own discipline

Challenges

- Slow coordination across functions due to departmental silos
- Decisions sometimes made far from the customer
- Lack of accountability for business results within functions

Matrix

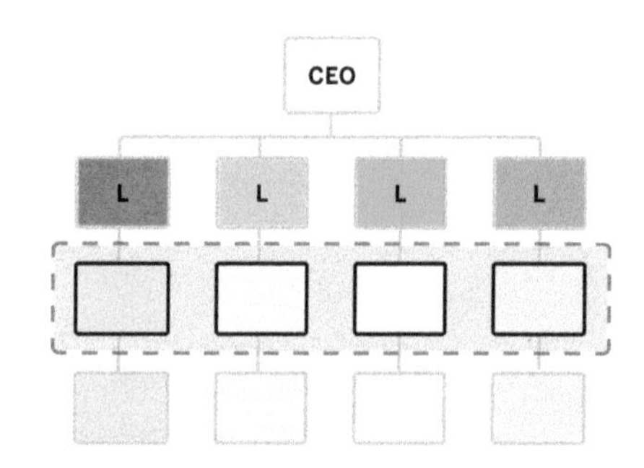

Examples

Philips's Consumer Health division uses cross-functional product teams for oral healthcare, mother and child care, and respiratory care.

Google's different functions come together to work on a specific product or project.

Key Stakeholders

Leaders of individual functions (primarily) and cross-functional team leaders (secondarily)

If you ask an engineer for a feature

"Let me check with my product manager after I finish this other thing for my manager."

Key Metrics

Focus primarily on function, e.g., cost per lead; secondarily on product, e.g., user engagement.

Benefits

- Enhanced coordination across functions
- More speed and autonomy for cross-functional teams
- More focus on business results at the team level

Challenges

- Competing priorities between functional managers and cross-functional leaders
- Potential duplication of effort and incompatibility across teams
- More stakeholders to manage

L Your primary leader

Your work group

Value Stream

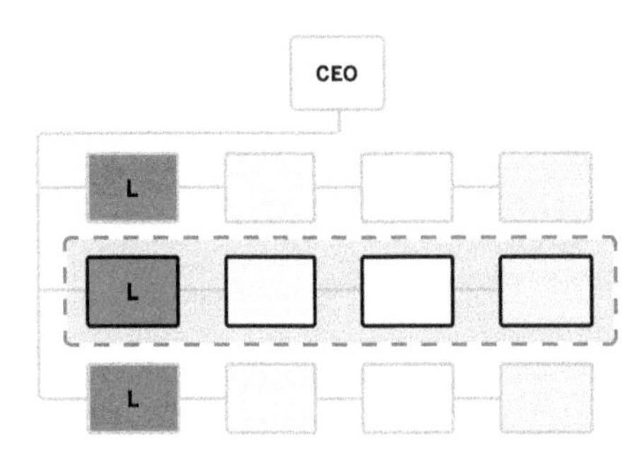

Examples

Spotify popularized cross-functional teams called "squads" responsible for product parts, with "chapters" and "guilds" supporting functions like engineering, or areas of interest like Javascript.

Hubspot product lines are led by product VPs and GMs, e.g., Marketing Hub and Content Hub.

Key Stakeholders

Cross-functional team leaders (primarily) and functional leaders (secondarily)

If you ask an engineer for a feature

"Let me check with my manager after I finish this other thing for my product manager."

Key Metrics

Focus primarily on results for the product, e.g., user engagement; secondarily on function, e.g., cost per lead.

Benefits

- Maximum speed and autonomy for cross-functional teams
- Closest coordination across functions
- Clear accountability and business results at the team level

Challenges

- Competing priorities between department managers and cross-functional leaders
- Greatest potential for duplication of effort and product incompatibility
- More stakeholders to manage

Divisional

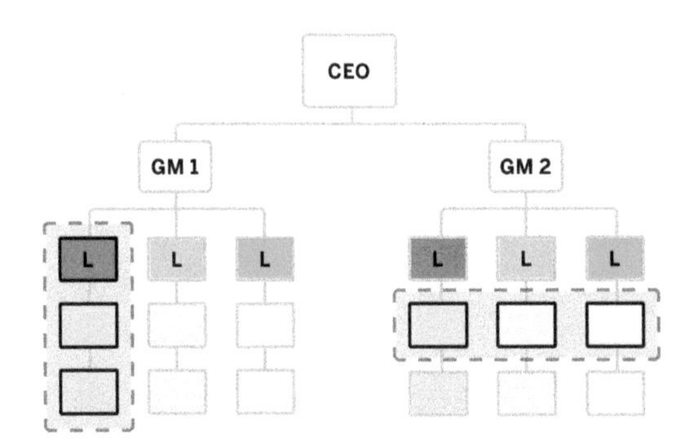

Examples

General Electric is organized into lines of business such as Aerospace, Digital, and Power.

Amazon is organized into divisions including Retail, AWS, Prime Video, and Devices.

Key Stakeholders

Leaders within your division according to that division's structure

If you ask an engineer for a feature

One of the answers from the other types, depending on the structure of your division.

Key Metrics

Focus on business metrics for the division, e.g., profit and loss or return on capital.

Benefits

- Focus on a particular market segment, industry, or line of business
- More autonomy for each division to organize differently
- Exclusive accountability for business results

Challenges

- Competition between divisions for resources, even for customers
- Duplication of resources and effort, product incompatibility
- Higher overhead costs for corporate leadership

Some companies follow these structures faithfully, but most take a hybrid approach. For example, a Functional organization may form a temporary cross-functional team for a strategic or experimental project. A large company with multiple divisions may use a different organizational structure in each division. Or the product, engineering, and design functions may work together in a Matrix structure, while sales and marketing operate separately.

Determine Org Structure and Key Stakeholders

A good way to determine the structure of your organization is to assess who gives direction (Figure 1-4). For example, you may be in a Functional organization if you, your engineering peer, and your design peer each get direction separately from leaders in your own function, i.e., product gets direction from product, engineering gets direction from engineering, etc. On the other hand, if you each get direction from your functional leaders, but a product manager or other cross-functional leader also provides direction to the Product Team,* then you may be in a Matrix. Who provides the final word between these leaders will tell you which stakeholders to prioritize.

Figure 1-4. Priority stakeholders based on who gives direction

Organizational structure	Priority stakeholders
Functional	Functional leaders
Matrix	Functional leaders (followed by cross-functional leaders)
Value Stream	Cross-functional leaders (followed by functional leaders)
Divisional	Based on the structure of your division

* The Product Team is the product manager, the designer, and the engineers who make up the core team working on the product. Some Product Teams contain other functions like user research or product analytics. In this book, we simplify the team as just product management, engineering, and design, but we recognize that your Product Team might look different.

Irie Determines Key Stakeholders

After Sri and Irie review the different types of organizational structures, they discuss what type of structure Helthex uses. "It seems like a Functional org structure," says Irie, "because the different functions each report to their own functional leader. It's definitely not a Value Stream structure."

"That sounds right," says Sri. "But my team acts like a Matrix. We have cross-functional Product Teams for each part of the product, where each team acts like its own unit. Everyone officially reports to someone of their own function, but they also have a dotted line to the leader of the team, who is usually a PM."

"Right," says Irie. "So Helthex is mostly Functional, but it's a Matrix in our org."

"Okay," says Sri, "so how does that help us understand who on the org chart are the key stakeholders for product?"

Irie suggests they go through the org chart and make a list of key players who, based on their position and responsibilities, should have input into product decisions.

Figure 1-5. The Helthex org chart

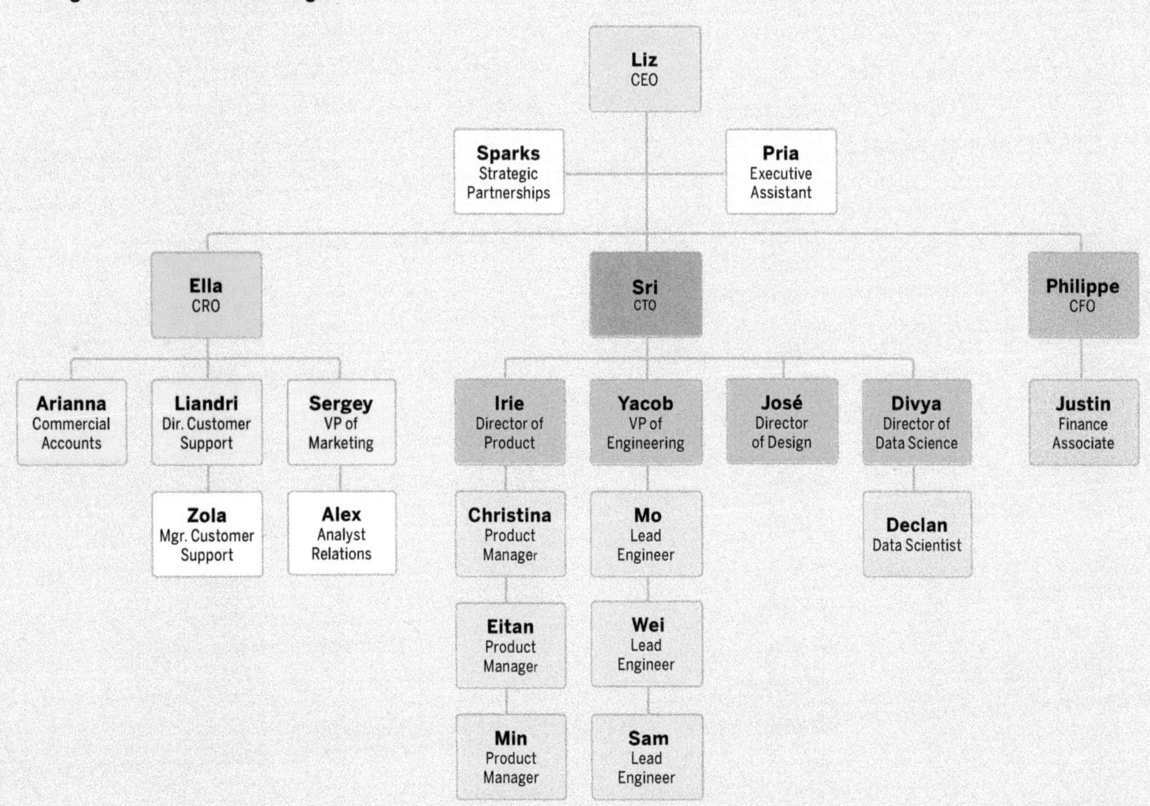

As she looks closer at the org chart, it confirms her suspicion that Ella, the chief revenue officer, has marketing (led by Sergey) and commercial accounts (Arianna), which sounds to her like some form of sales. Ella also apparently has customer support, led by Liandri. She recalls that Ella was at the product review meeting.

She isn't sure why a B2C company needs a commercial accounts person, but she sets that aside for the moment, moving on to Finance. She hasn't met Philippe, the CFO, but as with marketing, sales, and customer support, she is used to collaborating with finance on things like margin targets and investment criteria.

"I also need to sit down with Yacob and José to find out why we're prioritizing design enhancements right now," Irie says. "I assume there's a good reason, but I want to understand the business rationale. Then I'll want to meet Divya, since she's also on your team." Irie thinks for a minute. "I guess the people in the meeting this morning are the ones I should start with, because they are heads of departments, and are clearly interested in the product. Plus Liz, although I know she's out right now."

"Makes sense to me," says Sri as Irie writes her initial list. "You'll cover the heads of each function across the company, plus the functional leaders in our organization's Matrix structure."

"How do all these people interact with each other?" asks Irie. "Is there some existing meeting where strategy gets hashed out?"

Sri explains that the executive team meets every other week. "It's pretty tactical, though," he says. "Budgets, programs, sales and marketing numbers."

"What about product strategy?" Irie asks.

"That group shows up at our product review meetings, but there isn't much discussion of product strategy outside of that."

"I'm hoping I can change that," says Irie. ■

Figure 1-6. Irie's priority stakeholders list

	A	B
1	Name	Role
2	**Yacob**	VP of engineering
3	**José**	Director of design
4	**Sparks**	Partnerships
5	**Ella**	CRO
6	**Sergey**	VP marketing
7	**Divya**	Director of data science
8	**Philippe**	CFO
9	**Liandri**	Director of customer support
10	**Liz**	CEO

1.2 Irie Identifies Power Players

After Irie builds her stakeholder list, she looks back at the org chart and brings the conversation back to Sparks. "I'm still not sure how Sparks fits in," she states.

"What do you mean?" asks Sri.

"Well," says Irie, "he's not a functional leader because he doesn't have anyone reporting to him."

"Right," acknowledges Sri.

"But he's also not a cross-functional leader because he tends to work alone."

"True," says Sri.

"So what's his role?" asks Irie.

"Partnerships?" offers Sri.

"I think there's more to this org chart than meets the eye," says Irie. "I feel like I need X-ray vision to figure out what's really going on inside of it."

Sri laughs at this as Irie continues. "Actually, my old manager, Darius, was a master at this. He always seemed to know just whose ear to whisper into and what to say to get things done. Would you mind if I—if we called him to get his perspective?"

"You want to call your old boss and ask him about our company?" asks Sri, sounding dubious.

Irie laughs. "I guess that does sound weird," she says, "but he's an advisor now and he was always really helpful when I worked for him."*

Sri is still listening, so she continues. "I trust him enough that I'd probably call him for advice anyway, but I'd rather do it together. So we can both evaluate whatever advice he has."

"Well, that's very frank," Sri says with a sudden smile. "Why not? It never hurts to get some free advice."

Irie texts Darius and he replies that he's available the next day. They set up a time.

* We highly recommend asking for help when needed, from your manager, a peer, or even a third-party advisor or coach. Asking for help is better than wasting time struggling with something on your own ad nauseam. Getting other people's perspectives and learning new things makes you better at your job in the long run.

The next day, Irie and Sri call Darius from a small conference room. Irie explains the situation with Sparks. "You know that I'm good at building relationships with functional leaders, even C-suite folks," she begins. "But I'm not sure what to do with someone in charge of 'partnerships' with no reports and no team."

"Well," says Darius, "behind the official org chart, there's always a hidden org chart I like to call the 'influence org chart.' It's not a chart that you can draw out, per se, it's more like a list of the 'Power Players.'"

"Power Players?" asks Sri.

Darius explains, "These are the people who wield influence in the company, with or without the title."

"Like Sparks," offers Irie.

"Exactly," says Darius. "While your Sparks may not have a team or an obvious function himself, people listen when he speaks. This means he has influence over their priorities—or even their objectives—despite his unusual title."

"That's exactly how Sparks operates," Sri confirms. "He shows up in meetings—invited or not—and tells people what he thinks they should do." Sri adds a helpless shrug as he continues. "We mostly go along."

"Sparks is really just one example," adds Darius. "Power and influence are operating in every organization all the time, regardless of the official org chart. You said your group within Helthex is organized into a Matrix structure, Sri? Well, the influence org chart is what's really happening behind the Matrix."*

Irie and Sri listen as Darius describes Power Players and how to drive alignment with them on objectives and priorities. ■

* Whether seeing the world behind the Matrix requires the red pill or the blue pill, we leave as an exercise for the reader.

Power Players

Power Players are the people who can make or break your product because they have the personal authority to insist you take a particular direction or make specific changes. They can also be the people who approve your budget or enforce your adherence to legal, security, or other policies. Despite this level of influence, Power Players may not have all the context necessary to make optimal decisions, which can be a dangerous combination. These are the people who have the authority to execute a "swoop and poop" where they swoop in and tell you to completely change your direction at the 11th hour without fully understanding the consequences of their actions.

Many executives are Power Players, and the "reporting org chart" makes it clear who these are. But there are two additional types of Power Players who are often hidden on the official org chart: people within a "dominant function" and "CEO-Whisperers." It is important to identify these stakeholders because—just like the executives—they also have the ability to dramatically affect your product. You can only develop relationships with these stakeholders if you know who they are.

Dominant-function Power Players

These people get their power from belonging to the function in the organization that wields the most influence. If your organization is engineering-driven, for example, just being an engineer may mean you feel empowered to offer your opinion on matters outside of your function—on, say, the placement of a button or the price of a product add-on.

Use the questions on the following page (Figure 1-7) as a guide to determining the dominant function in your organization. You should be able to use a combination of observation and detective work to answer most of them.

Not all of your answers will point to the same place, but there will likely be a function that shows up in more of the answers than others. That is where you should start looking for the most influential stakeholders.

For example, imagine a company where the chief marketing officer worked at an advertising agency with the CEO in the past, they have lunch together most days, and most new initiatives are communicated and led by marketing. It would probably make sense to form a relationship with your counterpart in marketing to establish where your goals overlap.

Figure 1-7. Questions for dominant functions

Here are some more examples of evidence that points to a particular dominant function (Figure 1-8).

Figure 1-8. Evidence for dominant functions

Evidence	Dominant function
Saving a deal or a renewal usually wins over sticking to the roadmap.	**Sales**
Number one company objective is to grow revenue from new accounts.	**Sales**
Number one company objective is about cutting costs or margin.	**Finance**
Improving customer experience is discussed in most cross-functional meetings.	**Design or product**
New patents cause visible celebration.	**Engineering or legal**
Focus is on process efficiency and scalability.	**Operations or manufacturing**
All products must go through a compliance, risk, and/or security review before launch.	**Legal or compliance**
There is a "technical" roadmap and strategy alongside the product roadmap and strategy.	**Engineering**
The roadmap tends to get blown up when a big customer has a request.	**Customer support or account management**

CEO-Whisperer Power Players

Sometimes there is a key player with an innocuous title* who has great influence over the CEO. Building a strong relationship with them will help you succeed in your organization. Consider these questions when looking for a CEO-Whisperer.

Figure 1-9. Questions to identify CEO-Whisperers

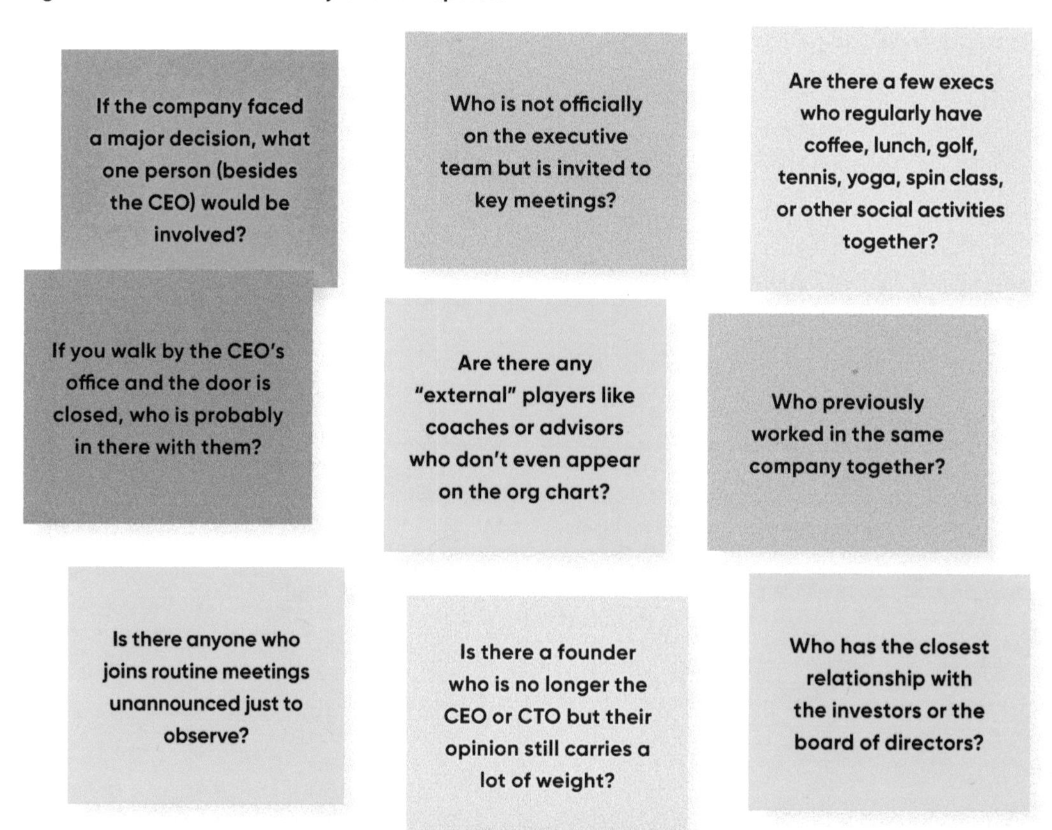

Consider the informal influence this person could have on a broad spectrum of your stakeholders. Bruce once reported to a charismatic CEO. Over time, though, Bruce realized that he needed the support of the CFO, a hidden Power Player, for most decisions. When Bruce wrote up a detailed pricing proposal and dropped by the CFO's office to discuss it, the decision to move forward was made in 10 minutes. With the CFO's buy-in, everyone else fell in line, including the CEO.

---

* Like "Strategic Partnerships."

Upstream Versus Downstream Stakeholders

Upstream stakeholders help bring a product to market, while downstream stakeholders market, sell, support, or use the product after it is released. It is important to collaborate with both types of stakeholders from the beginning because both are important to the product's ultimate success.

Obvious examples of **upstream stakeholders** are members of your Product Team, such as engineers and designers. In addition, you likely depend on teams that help you collect or analyze data, like marketing or data analytics. Teams like legal or data security may advise you on policy or regulations, which can be inputs into your strategy. You may also need people operations to help you hire and build out your team. Another type of stakeholder you'll likely encounter are "gatekeepers" who may need to approve your product or some aspect of it, such as security auditors or a pricing committee.

Downstream stakeholders include customers, internal stakeholders who support customers, and those who market, sell, and support the product. Your primary downstream stakeholders are marketing and sales, as they will be critical in getting the product into the hands of customers after it's released. Also valuable are customer support, customer success, and training teams because they can help represent customer needs from a learning or onboarding perspective. Customers themselves, while not in the scope of this book, are also important downstream stakeholders as the key users of the product.

Moving Power Players Toward Alignment

Once you have identified your Power Players, what do you do next? Aligning with your Power Player stakeholders is critical for the success of your product because—without that alignment—they have the potential to be extremely disruptive to your roadmap. So your next step is to figure out which Power Players are aligned with your goals and which are not.

Someone is aligned with a decision or with overall product direction if they see value in what you are doing, actively want you and your product to succeed, and will help you to accomplish your objectives. They may have suggestions (sometimes a lot of them) and they may not agree with every decision you make, but aligned stakeholders are willing to go along and try their best to make things work.

Power and alignment can be mapped simply, as in Figure 1-10. Plotting your stakeholders in this way can help you identify stakeholders who present the greatest risk to your product, and then prioritize moving them toward alignment. See Figure 1-11 for how to prioritize among people in these categories.

Figure 1-10. Power versus alignment

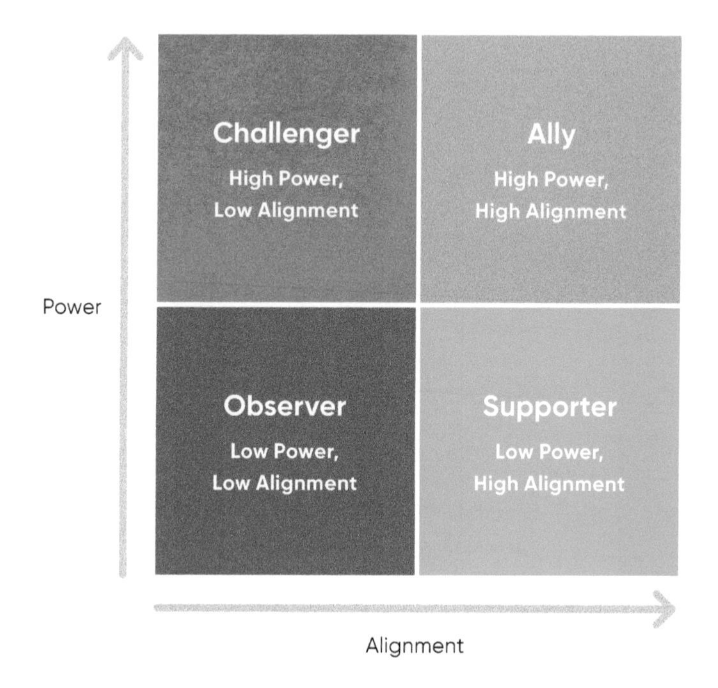

Figure 1-11. Prioritizing stakeholders

Quadrant	Description	Examples	How to Prioritize
Ally **High Power, High Alignment**	Power Players or people actively contributing to your product who are invested in its success and supportive of your plans	Your product's executive sponsor, members of your Product Team (hopefully)	Actively maintain alignment with this influential group
Challenger **High Power, Low Alignment (Danger)**	Power Players who are indifferent to your product plans or actively opposed to them	An executive with a lot of ideas that don't fit with your plans, a Product Team competing for the same budget or people, an executive fighting with your boss	Use every effort to influence this potentially dangerous group toward alignment
Supporter **Low Power, High Alignment**	People invested in your product's success, and supportive of your plans, who have little power to influence your decisions or your alignment with others	An individual customer support rep or salesperson with few or no complaints	Leverage these interested folks for ideas and feedback on new product initiatives
Observer **Low Power, Low Alignment**	Anyone indifferent or even opposed to your product plans but who has little power to help or hinder you	An account manager for an unhappy but very small customer, people in other divisions, or those working on noncompeting products	Worry about this low-priority category last, if at all. If you think they could contribute in some small way, consider a low-key awareness campaign such as demoing your product at an all-hands meeting or sending out a monthly newsletter.

Ideally, all of your stakeholders would be completely aligned all the time, rallying around your product and praising your brilliant plans. This is rarely the case, however, so your job is to maintain alignment where you have it, and to move unaligned stakeholders toward alignment. It's not practical to actively engage with everyone all the time, so you should begin by focusing on the most powerful.

Following are detailed descriptions of two key ways to move any stakeholder from Challenger to Ally, or from Observer to Supporter: tie your product's success to their own objectives and invite them into your process.

Tie your product to your Power Player's success

A Power Player whose interests are not aligned with yours is dangerous. They can hamper your progress merely by diverting resources to other initiatives. By actively opposing you, they can send your product to an early grave.

To gain their support, establish a link between your work and their focus. If the CMO wants to launch a big marketing campaign, pitch your new release as the ideal opportunity. If the CFO is focused on cost-cutting, showcase the design enhancements you think will reduce support costs or the performance improvements that will drive down server utilization. If the CEO has created a new business strategy, get them to support your new initiative by showing how it supports that strategy. Gain alignment by showing them how your work will make their plans successful.

Melissa worked at a supply chain company where the COO created a new strategy pillar around reducing the number of damaged items that customers received. At the time, Melissa was working on an initiative to streamline the fulfillment centers by systematically directing the work that associates performed. Because a major goal of her initiative was damage reduction, she was able to gain the COO's support for her streamlining initiative by framing it as the foundation for the new damage-reduction strategy pillar.

Invite Power Players into your process

If the direct approach is not working, consider asking for their ideas. By asking for a Power Player's advice, you are satisfying their need to exercise influence while simultaneously retaining at least some power for yourself to finalize decisions. Most people are flattered to be asked what they would do in your position, and it engages their empathy, making them want you to succeed.

When you find the CEO-Whisperers in your company, approach them as respected authorities, even if their title doesn't match their influence. Ask informally for their advice and guidance. Words like "I am working on my plan and I could use your perspective" can be disarmingly effective.

Sometimes asking for approval directly or showing too much subservience can backfire. People may be embarrassed if they think you are trying too hard to win their approval. With people like this, ask if they "have any feedback" on your plan rather than "is this okay with you?" Each culture and company is different, though, so use your judgment or ask a trusted colleague for advice on how to approach specific Power Players and what wording to use.

Whether they are Power Players or not, when you move people toward the upper right of the Power/Alignment grid, you are essentially recruiting them to your team. They may not be a formal part of your Product Team, but if you keep them informed, ask their advice, and treat them with respect, they will tend to return the favor.

Bruce once had the tough job of retiring a product that would soon become unprofitable, but that was still producing revenue. As a product manager, Bruce worked with finance to demonstrate that, as suspected, technical updates that were required by regulations would make the product wildly unprofitable. Bruce suggested to the head of finance that he take this project as an opportunity to showcase the skills of his team.

Bruce also worked with sales to develop a plan for replacing the lost revenue, with the legal team to ensure the company wouldn't violate any contracts or license agreements, and with support to provide options for customers still using the product. These folks joined Bruce's temporary team and they worked together for a few weeks. Once the plan was approved, everyone went back to their regular duties.

Irie Identifies Sparks as a Power Player

After the call with Darius, Irie and Sri agree that the official org chart doesn't represent who has real power within the organization to influence product decisions. They decide to map the key players at Helthex according to the Power/Alignment grid and add this information to Irie's stakeholder spreadsheet.

Irie projects her laptop on the screen so Sri can see it too.

"Let's start with the people in the upper right: Ally," says Irie. "If I'm thinking about my role as director of product management, I'm going to put you, Yacob, and José in there." Irie notes this on her Stakeholder Canvas. "If I were an individual contributor, I'd put my Product Team in there, including the engineering lead and all the engineers, and the designer, if we have one dedicated to our team."

Sri smiles. "I suppose if your own team isn't aligned, you have bigger problems."

"Exactly," says Irie. "For right now, I'm going to put Ella in the upper left: Challenger. From what I saw in the product update meeting, she has a lot of influence on the roadmap and she is not aligned with the Product Team's priorities, or even with other stakeholders."

"I hate to admit it," says Sri, "but I think you're right on that one."

"What about Liandri?" Irie asks.

Sri relates that Liandri maintains a list of support issues she'd like to see addressed but that it doesn't get as much attention as it probably should. "Ella, in sales, and Sparks tend to overrule her," he says. "But I think you'll find her team has a pretty good sense of what's important to customers. She may not be aligned with the others, but I'd say she is pretty well aligned with our priorities." They agree to place Liandri in the Supporter category.

"Sergey is probably in the Observer category," adds Irie. "He seems to be not too involved in the product, except when my team asks his team for help on a product launch."

"That's accurate," says Sri. "You haven't met with Divya one-on-one yet, right?" he then asks. "She can sometimes be challenging. She's super smart and can occasionally be overly direct, but I think she's basically aligned on goals and direction." Irie records a question mark for Divya pending a one-on-one session with her.

"So what about Sparks?" Sri asks.

"What about Sparks, indeed?" asks Irie. She thinks about it for a minute, then asks, "Do you think he's in the same bucket as the C-suite, even though he doesn't have a C in his title?"

"Yeah, I'd say so," says Sri. "Next to Liz, he's really the biggest Power Player. I'm hoping we can get to alignment with him eventually, but Sparks should start in the 'danger' category in the upper left."

Irie agrees and updates her stakeholder list.

"This is a good start," says Irie. "And it's a helpful map of where I need to work on alignment. Knowing that Sparks is High Power but

Low Alignment makes it clear where I need to focus."

Shifting gears, Irie says, "I'm not sure how to approach him, though. He seems to want to just dictate features. That's not how I am used to operating. Maybe if I knew more about what he is really trying to accomplish, we could have better conversations. If I understood his goals better, we could discuss feature ideas in that context."

"Should feature work be dictated by Sparks's goals?" asks Sri.

"That's a fair question," Irie responds. "We should really be driving product strategy from company strategy and objectives. In Liz's absence, maybe I can start by asking him about those and then about his own goals."

"That sounds like a good plan," says Sri. "We don't talk about this stuff enough." ■

Figure 1-12. Irie's stakeholder list updated with Power/Alignment categories

	A	B	
1	Name	Role	Power/Alignment
2	**Sri**	CTO	Ally
3	**Yacob**	VP of engineering	Ally
4	**José**	Director of design	Ally
5	**Sparks**	Partnerships	Challenger
6	**Ella**	CRO	Challenger
7	**Sergey**	VP marketing	Observer
8	**Divya**	Director of data science	?
9	**Philippe**	CFO	?
10	**Liandri**	Director of customer support	Supporter
11	**Liz**	CEO	?

1.3 Irie Learns About Decision Making

Irie, now in week two at Helthex, feels like she's starting to get into the swing of things. She's met individually with each of the three people who report to her (Christina, Eitan, and Min), and she attended the joint "Product/Design Weekly" meeting with those two teams. Yacob (vice president of engineering) and José (director of design) took her out to lunch in her first week, and she joined the engineering team's virtual happy hour on Friday afternoon. She also attended each Product Team's sprint review, which gave her a sense of what the teams are working on.

No one from outside of Sri's org has reached out to her, and this reinforces her impression of the overall Functional org structure. She has been in a few meetings with Sparks, but she can't quite get a read on what motivates him. He asks for a lot but he seldom explains his reasoning. People aren't sure whether or not to accommodate him or who should be involved in making that decision.

Irie remembers that Eitan said something about the decision-making culture in the product/design team meeting, so she asks him to help her understand how decisions are made at Helthex.

Eitan appears on Irie's laptop, and he's a bit blurry, like he has a bad connection. His virtual background, though, is crystal clear. "Hi Irie," says Eitan, "you wanted to talk about decision making?"

"Yes," says Irie. "Thanks for taking the time to talk with me."

"Of course," says Eitan. "I think you can make a big difference here and I want to help you get a solid footing."

Irie smiles. "I'm interested in hearing more about how you make product decisions, where you get ideas from, and who in the company is most influential."

"Well," begins Eitan, "some of our ideas come from customer research: discovery interviews and user testing. José's researchers are amazing with that stuff." He pauses. "But to be honest, we get so many internal requests it's hard to keep up."

"Where do those requests normally come from?" asks Irie.

"I get a lot of requests from sales and support," Eitan begins. "Arianna's sales team puts in a lot of tickets, and I'm never really sure how to prioritize them." Eitan looks downward as he finishes, his voice growing quiet.

"Your team handles admin features, right?" Irie asks. When Eitan nods, she asks, "You get a lot of requests from sales for administrative features?"

"Support asks for internal capabilities like bulk-loading history data from another app," he clarifies. "But sales has endless requests for things that B2B customers ask for. They pass all of those directly to us."

Irie adds Arianna to her stakeholder list, annotating her as "High Power," but leaving the alignment dimension blank for now. She then looks back at Eitan and asks, "How do you decide what to work on in the end?"

Eitan explains that he maintains a massive spreadsheet with all the tickets that people have submitted from around the company. "For a while, I got representatives from all the functions to vote on the things they thought were most important, but it was hard to keep it up. Unfortunately, we end up working on what people shout about the most, which isn't a great method." Eitan continues, with obvious frustration. "There's just no way to please everyone, so I end up doing little bits every release from everyone's agenda. I feel like we never really nail anything because we can't focus."

"I like the way you tried to include people from around the company," Irie says. "I believe making a product successful is a team effort—and everyone should have an opportunity to give input. I don't think that majority rule—or rule by the loudest, for that matter—is the right way to leverage that input, though."

"I know," says Eitan. "I don't like it either. So I decided to start reading about how decisions are made at other companies, and I came across a model from Bain & Company about decision-making cultures, which is what I was describing in the product/design weekly meeting."

"Interesting," says Irie. "Tell me more." ■

The Four Types of Decision-Making Cultures

To influence decisions, you must first understand how decisions are made, because decision-making culture will vary by organization and even by department. The consulting firm Bain & Company describes four organizational decision-making cultures: Directive, Democratic, Participative, and Consensus.* We will explain these here in the context of product management (Figure 1-13).

Figure 1-13. Four decision-making cultures

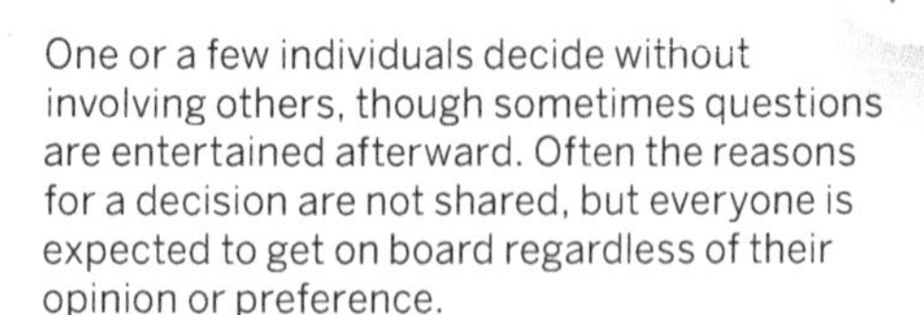

Directive

One or a few individuals decide without involving others, though sometimes questions are entertained afterward. Often the reasons for a decision are not shared, but everyone is expected to get on board regardless of their opinion or preference.

Benefits

It's easy to understand who makes the decisions because it's always the same person/people.

This Is Your Org If:

Decisions are handed down from upper management without much input.

Challenges

Decisions are often made without all the necessary inputs or information needed, which may cause unintended consequences.

Key Stakeholders

Support from a short list of Power Players is critical. No one else's opinion really matters (unless they can influence Power Players).

Make It More Participative

Advocate for involving knowledgeable stakeholders to improve decision quality.

Example

An eccentric billionaire buys a company, fires half the staff, and tells the remaining employees they must commit to the new vision but doesn't explain what that is or how it will be achieved.†

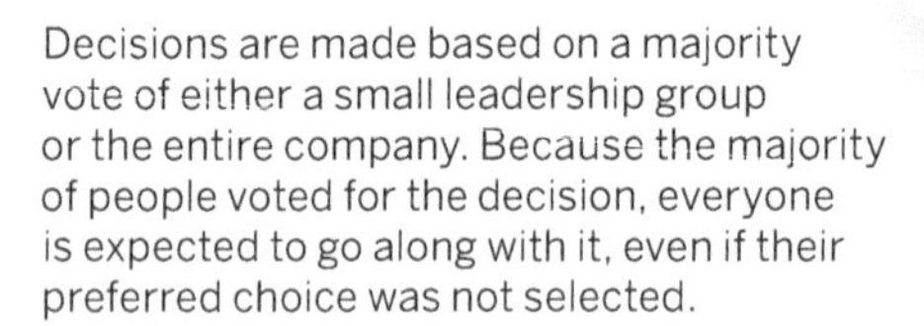

Democratic

Decisions are made based on a majority vote of either a small leadership group or the entire company. Because the majority of people voted for the decision, everyone is expected to go along with it, even if their preferred choice was not selected.

Benefits

Everyone in the voting group gets an equal say in decisions, improving buy-in.

This Is Your Org If:

Decisions are made by taking a vote and going with the majority choice.

Challenges

The majority may not favor a necessary shift in strategy, leading to suboptimal decisions. There may not be a clear decision maker to break ties.

Key Stakeholders

Align with Power Players who can influence the votes of others or who can determine which decisions get on the ballot.

Make It More Participative

Advocate for a single person to communicate context, lay out options, and organize the vote.

Example

A working group comes up with three different ideas for a new name for the company's flagship product. They present the three choices in an all-hands meeting, and everyone votes.

* Marcia W. Blenko et al., "Shape your company's decision style—and behaviors," May 11, 2011. *https://www.bain.com/insights/decision-insights-8-shape-your-companys-decision-style-and-behaviors.*

† Also, they change the name of the company to a single letter. That would never happen in real life, of course.

Participative

One person is designated as the decision maker for each decision or type of decision. That person is expected to actively seek and consider input from stakeholders who have relevant expertise or interest. Everyone is expected to commit to the decision maker's choice.

Benefits

Combines the speed of single-person decisions with the quality and buy-in from broad inputs.

This Is Your Org If:

Decision makers are clearly identified at all levels of the organization.

Challenges

Failing to consult with Power Players and others with strong opinions may invite criticism and second-guessing later.

Key Stakeholders

When you are the owner of the decision, you must determine the list of people to consult, which should include Power Players.

Make It More Participative

Advocate for clearly identifying a decision maker early and choosing good input providers.

Example

A product manager decides to enter a new market after consulting with marketing, sales, engineering, and finance. Everyone commits to making it work, despite some disagreements.

Consensus

Nearly everyone is involved in making the decision. Without a clear decision maker or a decision-making mechanism, decisions tend to get stalled in extensive discussion and debate, and are frequently revisited or reversed. Often decisions are never made at all.

Benefits

Decisions are not made until broad agreement is achieved, improving buy-in.

This Is Your Org If:

Decisions are not made until everyone agrees, so sometimes decisions are never made.

Challenges

Decisions are made slowly and often require multiple compromises or concessions, reducing the effectiveness of the eventual solution.

Key Stakeholders

Everyone is your stakeholder because you need to get all parties on board with any decision that needs to be made.

Make It More Participative

Advocate for a single person to collect input and negotiate with individuals to align on a decision.

Example

Design wants better onboarding, engineering wants less tech debt, sales wants competitive features, and support wants bugs fixed. The PM includes a bit of each to appease everyone.

Different situations use different decision-making approaches

Within an organization, there may be different decision-making cultures in different departments. Decisions also may be made differently based on the type of decision being made. For example, whether to have tacos or pizza for a company event may be a directive decision by the organizer. That same organizer might put the location for the event to a vote, however, and might postpone a decision on the agenda until a consensus is reached on the organizing team. It's the pattern of decisions over time that will tell you about the dominant decision-making culture of the organization, and this informs the expectations people have.

The decision-making culture of an organization can also change under different circumstances. In *The Hard Thing About Hard Things* (Harper Business, 2014), Ben Horowitz argues that a CEO must take a more Directive approach than usual when the company they lead faces an existential threat. He writes, "Peacetime CEO knows that proper protocol leads to winning. Wartime CEO violates protocol in order to win."

Participative is best for product management

Product decisions are made quicker if there is a single person responsible. The Participative approach ensures better decisions than a Directive one because not only does it have a single decision maker, but it also incorporates input from multiple sources. That cross-functional input tends to drive the buy-in needed for strong execution. Bain & Company says the Participative approach "combines single-point accountability with a collaborative approach to the process."*

We've found that the Participative decision-making culture is typical of healthy cross-functional Product Teams found in the most successful tech companies. Product managers in these teams are ultimately responsible for product decisions, like what features to develop, but they work closely with engineers, designers, marketers, salespeople, and others to make those decisions.

Because different types of decisions can be made differently, you can influence the way product decisions are made without having to change the decision-making culture of the entire company.

---

* *https://www.bain.com/insights/decision-insights-8-shape-your-companys-decision-style-and-behaviors*

Irie Identifies the Decision-Making Culture at Helthex

After Eitan finishes explaining the four decision-making cultures, Irie uses what Eitan said earlier about decision making at Helthex to suppose that Helthex uses a mix of Directive and Democratic approaches.

"That sounds about right," says Eitan. "When there's a C-suite person or a director in the decision-making process, they usually make all the decisions. And when they're not there, people don't know what to do, so we vote on the decision. It doesn't seem like the best approach, to be honest."

"I agree," says Irie. "A Participative decision-making culture is what I'm used to. That's where the product manager makes the final decisions about the product, but with input from their team and stakeholders."

"That sounds a lot more efficient and effective. We wouldn't have to wait for a VP to make the decisions, and the voting method doesn't always give a good answer at the end."

Irie adds, "I seem to remember Sri telling me about Sparks wanting to expand into Canada as a new market, and everyone talked him out of it."

"I remember that," says Eitan. "That one sounds like Consensus."

"It sounds like we do everything except Participative," says Irie.

Later that day, Irie heads into her one-on-one with Sri with some questions. First she explains the basics from what Eitan shared about decision-making cultures.

"It sounds like Sparks primarily fosters a Directive decision-making culture. He likes to make all the decisions himself," observes Irie.

"I've used this Participative type of decision making at other companies," says Sri, "but not here. Here, Sparks keeps changing the strategy on us. We just get started on one thing and it changes with no warning or reason why. It's hard to make any progress or have any sense of control."

"Yeah, that doesn't sound good," Irie sympathizes. "I'd like to convince him that Participative is the way to go, but based on what I've seen so far, I'm worried that it will be a challenge."

"Since we know Sparks is a Power Player who's not yet aligned with our team, I think the best approach is to just be honest about the negative effects of the current process and how Participative decision making could work. Bring him into the process."

"I'll try that," says Irie. ■

1.4 Irie Determines Decision-Making Authority

As Irie enters week three, she finally gets to meet Divya, head of data science, in a technical brainstorming meeting with Sri and Yacob. She takes a few minutes at the end of this session to outline the Participative decision-making model, and Divya is immediately supportive.

"This is how I am used to working." Divya says. "My team should own data decisions, just like yours should own technical ones," she adds, turning to Yacob. "But we get overruled by Sparks at a whim. We even get argument from engineers on our models," she complains. "Your team is pretty good, Irie. Occasionally, we get pushback from them."

"Pushback?" asks Irie.

"Just last week," Divya explains, "Christina asked Declan on my team to add an additional input to the exercise readiness model. One of her data sources can give us a measure of how much physical work the person has performed in the last day and she thought it should go into the model."

"Makes sense to me," says Sri. "You guys know I track my workouts every day and I know I overdo it sometimes. Maybe this would keep me from overtraining."

Divya rolls her eyes. "You are making my point for me, Sri. You think this would be useful for the model because you don't understand how it works. The model is built on how your body reacts to stress. We can tell from heart rate variability and resting heart rate whether you are overtraining. How much work you did doesn't really matter. It's the effect that work has on your body that tells us whether you are ready to do more. You and Christina should leave these decisions to the experts."

Irie suggests they ask Christina to join them to explain her point of view. Divya objects, but Irie asks her to go along so they can play out the Participative model. "If you are not convinced, it's still your decision," she says, and Divya agrees.

Christina is working from home but she is able to join them on video. Without providing much context, Irie simply asks Christina what her reasoning was for proposing they add this variable to the model.

"I have to admit," Christina begins, "that it was my data partner who initially suggested it. They are making a push with this data and they thought it would be useful." Divya rolls her eyes again, but Christina continues. "I looked into feature requests and support inquiries, though, and it turns out lots of users are asking about it. Either they want to know if we use that data in the readiness model, or they know we don't and they want it incorporated."

"But the users aren't data scientists either!" exclaims Divya. "They shouldn't be designing our models any more than any of you!"

"Then," Christina continues, "I read a study that showed that models that included this variable were better at predicting risk of injury."

"What?" asks Divya, surprised.

Christina offers to forward the study to Divya. "I don't know if it's the right thing to do," she clarifies, "but I thought you guys should look into it. I tried to share the study with Declan, but he wouldn't consider it."

"Okay, sure," says Divya. "We can look into it. I didn't realize how much homework you had done."

"How would we ideally want a decision like this to work?" Irie asks the room.

"I have to give Christina credit here," Divya begins. "I think Declan and I didn't give her enough of an opportunity to offer input. But I feel we still need to retain the final decision."

"That's the essence of the Participative model," Irie says.

"Yes," replies Divya. "I think I was just focused on the single-owner part."

"Is Christina the only one who can give input to Declan on models he owns?" asks Yacob.

"You're asking for permission for engineers to comment on our models, too," says Divya, sounding reluctant. "I'm worried about this getting out of control."

"Shouldn't anyone be able to weigh in?" asks Sri.

"We're not designing models by committee!" insists Divya.

"No," says Irie, "that would be the Consensus model, or Democratic."

"I'm starting to think they are no different," says Divya, looking miserable.

"Most people won't have any input or even any interest," says Yacob. "But if one of my team wants to make a case, I think your team should at least listen. How is that not reasonable?"

"Let me try to summarize what I think you all are saying," says Irie. "I don't think they are so incompatible. Actually," she adds, "let me bring up this article on the DACI framework I found that pretty much describes it."

"Do you mean RACI?" asks Sri.

"Similar," says Irie. "RACI is for responsibilities in general. DACI is specifically for decisions. Reading this article the other day, I realized we'd been moving toward this framework without formalizing it." ■

The DACI Decision-Making Model

Group decisions inevitably bring up questions of accountability and decision-making authority. If you and I disagree, who decides? It's a bit meta,[*] but making a decision about who makes the decisions in different situations is critical to avoid confusion and delays.

DACI stands for "Driver, Approver, Contributors, Informed."[†] Use this model to determine who is involved in which decisions, and in what way. DACI is especially useful in a Participative decision-making culture, where decision authority isn't always clear. In fact, simply discussing the DACI model with stakeholders can actually help them move toward a Participative decision-making culture by helping clarify how decisions are made.

Figure 1-14 explains the four DACI roles with examples.

Figure 1-14. DACI roles

Role	Description	Example
Driver	Aligns with the Approver on goals and decision criteria, seeks input from Contributors, and makes the decisions	A product manager who decides what goes on the product roadmap
Approver	Sets decision criteria, approves or rejects decisions by the Driver	A VP of product who can reject (or inject) roadmap items
Contributors	Add valuable information to enhance the decision or clarify the impact	A customer service rep who proposes something for the roadmap based on customer challenges
Informed	Updated about decisions by the Driver but do not contribute	An accounts payable clerk who sees the roadmap for the first time when the PM presents it

* Like a coffee table book about coffee tables...

† DACI may sound similar to RACI (Responsible, Accountable, Consulted, Informed) but DACI is used for decision making, whereas RACI is used primarily to determine who has which role for doing the actual work.

How to use DACI roles

DACI roles are critical to create an efficient decision-making process. Once someone takes on the Driver role for a particular decision, it is their responsibility to initiate an iterative alignment process that involves the Approver, Contributors, and Informed stakeholders (Figure 1-15).

The DACI roles provide a structure for the Driver to make the Participative model work. By requiring alignment with the Approver on objectives, and input from Contributors on decision options, the decision quality improves and challenges to the decision drop dramatically. People are much more likely to commit to a direction if they feel they've contributed and been heard, even if they wouldn't have made that decision themselves.

Figure 1-15 DACI process

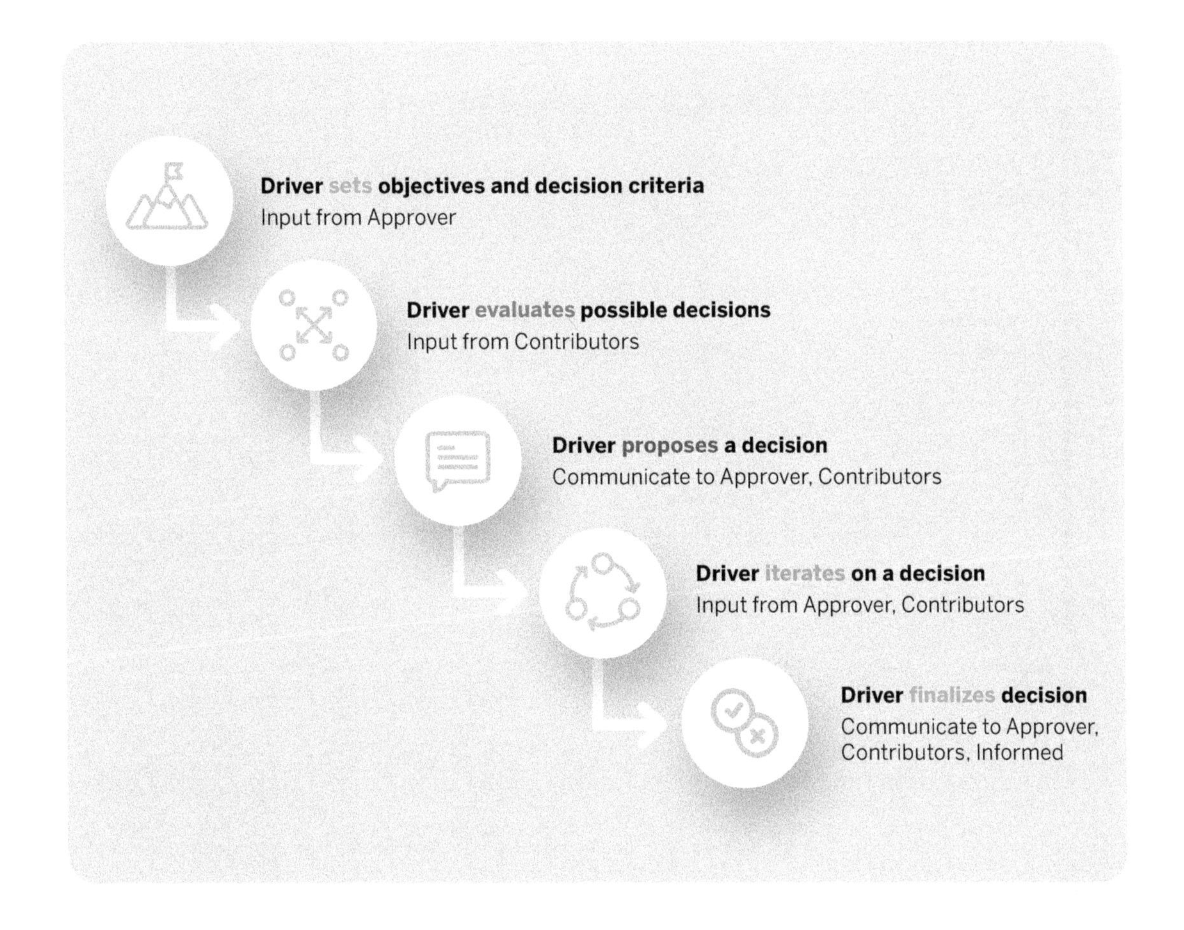

Stakeholder Canvas

Consolidating what you're learning about your organization and individual stakeholders can prevent it from becoming overwhelming. We recommend a simple tool we call the "Stakeholder Canvas" (Figure 1-16). This is a worksheet to organize key stakeholder insights for quick reference, such as their Power/Alignment category.*

Your Stakeholder Canvas is best kept private. Sharing it may create privacy issues for the individuals on the list, and they may not appreciate your judgments about them (even if they're true).

Figure 1-16. Blank Stakeholder Canvas

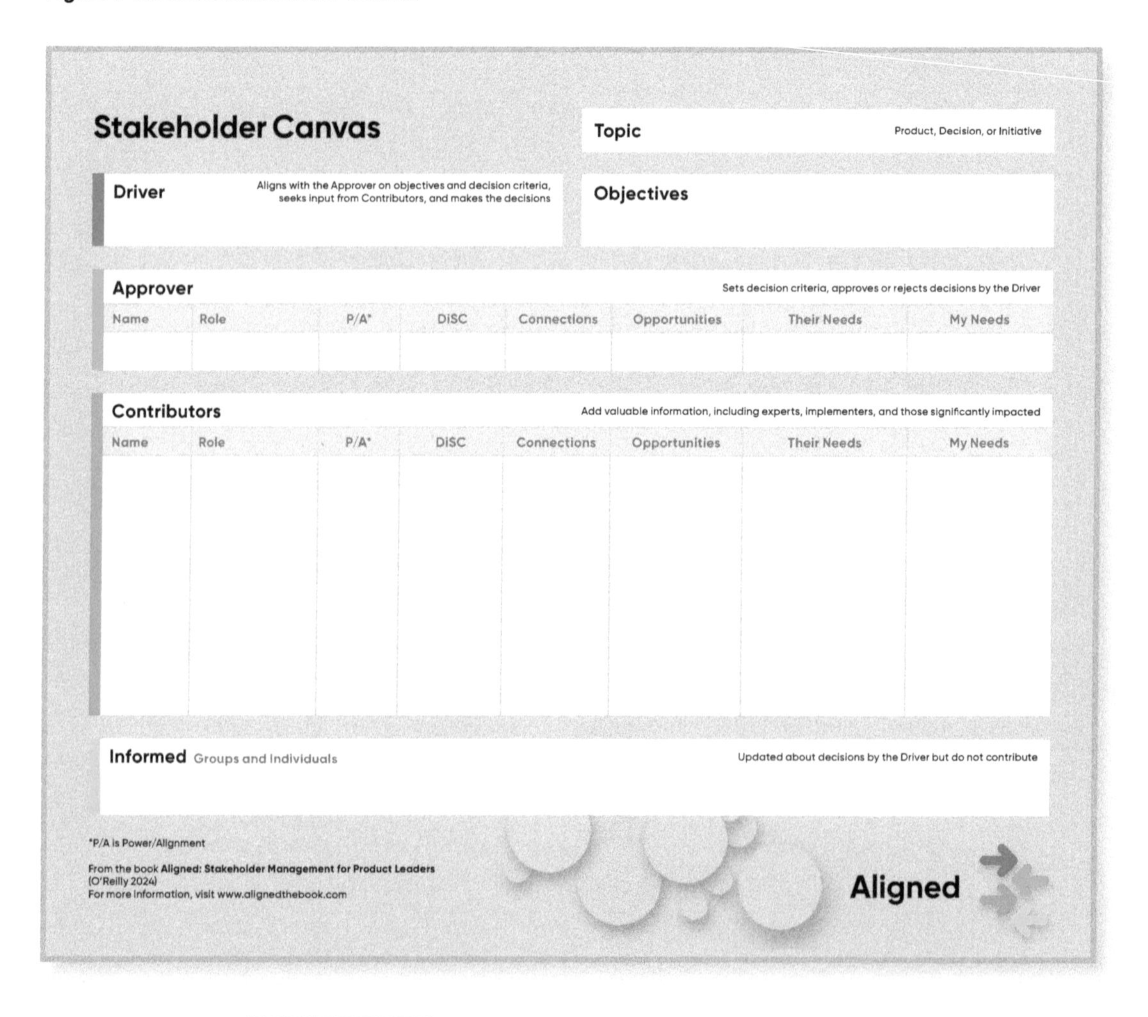

* You can find a downloadable Stakeholder Canvas template at our website: *alignedthebook.com*.

Irie Creates a Model for Participative Product Decisions

After walking through the DACI model, Divya summarizes how it might work for data science decisions, like the inputs to Helthex's exercise readiness model.

"Declan would be the Driver," Divya says. "He'd own the final decision as to what goes into the model. He'd agree with the Approver on criteria for the decision. The Approver would be me?" she asks.

"As the head of data science, yeah," agrees Irie. "You'd be able to overrule Declan if you think he made a mistake or didn't have all the context for the decision."

"Okay," says Divya, continuing. "Declan would seek input from Contributors. And you're saying that anybody could contribute if they want."

Sri chimes in. "I think Declan would actually seek out people who are likely to have useful input," he says.

"Like Christina with the data about customer inquiries and the study," offers Divya.

"Right," says Yacob. "Or, actually, some of my team does have experience with data science. Even if they didn't go to school for it, they've been implementing your models for years and seeing how things work in the real world."

Divya is nodding. "That is true," she admits. "I think pragmatically it would make sense to ask a tech lead about some options that consider performance."

Yacob says this is a good example and Divya summarizes. "So Declan would seek input from Contributors, develop a proposal, get feedback on that proposal from Contributors, and then finalize the decision." She then asks, "Do I have to approve it before it's official?"

"Most of the time, no," explains Irie. "If you've aligned on criteria, and he's got sufficient input, Declan can simply inform everyone of the decision. If you have to jump in at that point, something has gone wrong along the way."

"That matches up with how I run things with the tech leads," says Yacob. "We've worked out technical design principles, architectural approaches and such. The tech leads decide on the approach they'll take for implementing things within their team. If it's new territory, they ask me and each other for input, but after that, the decisions are up to them."

"Aren't you worried they'll make a mistake?" asks Divya.

"It's happened," Yacob admits. "Once Wei wanted to try a new approach to using one of the backend APIs. She didn't check with the team that maintains that API and she brought the whole system down with too many calls once we were in production. We had to roll the changes back and add in rate limiting to fix it."

"I remember that," says Sri. "We were down for half a day."

"From the sound of it, that was because Wei didn't get input from all the Contributors," Irie observes. "If she'd been using DACI, she would have thought about who to ask for input. The API team would have been an obvious candidate. It's possible that you would have caught it, too, as the Approver, Yacob, but the API team would definitely have asked questions in that case."

"Maybe the API team should be the Approver for something like that," Sri suggests.

Yacob shakes his head. "I don't want that team bogged down with every tiny change to how APIs are used. Plus, you can't assign a decision to a whole team—it has to be an individual."

"That would slow the teams down that use those APIs, too," says Christina, "if they had to wait for the okay on everything."

"I expect we can trust the teams enough to know when they need to get input," says Yacob. "We all learned that lesson after that outage—especially Wei."

The room is silent for a moment. Irie turns back to Divya to ask if she thinks this process can work for data science, too.

"Let's try it," she says. "But I'm going to raise the alarm if we try to turn this into a democracy."

"I'll be right there with you," says Yacob.

"Then we are all agreed," says Irie.

After the discussion about how the data science team should make decisions, Irie drafts a DACI for key product decisions, based on how she's seen it done in other companies.

She also updates her Stakeholder Canvas to include DACI roles for high-level product decisions, like changes to strategy. She adds Arianna, from her conversation with Eitan, and also adds some of her closer team members, who now have roles on the DACI map, including herself.

Satisfied that it's a good starting place, she brings her drafts with her to Sri's office and asks for his input.

"I think Liz would be an Approver, too," Sri says after a quick review.

Irie explains that for a given decision, there can only be one Driver and one Approver. "Otherwise it gets confusing as responsibilities are spread around."

Sri objects, pointing out that, as CEO, Liz can overrule anyone. "Yes," replies Irie, "but she's delegated product decisions to you as CTO, right? She could pull rank, but she can't do that frequently without you quitting, because then you'd have no real authority. This is the essence of the Participative decision style," she adds. "Each decision is driven by one person with one Approver and lots of Contributors."

Sri acknowledges this, then asks, "If we limit the Driver and Approver to single people, do we need this many Contributors?"

"We have the authority to make product decisions," says Irie, "but we can't exercise it in a vacuum. A company is like a machine for making money by solving customer problems. You can't change one part of that machine without adjusting other parts of it."

"I suppose," says Sri.

"And we need information," Irie continues. "If we're losing a lot of deals due to competitive gaps, I might be able to get that information from marketing. If customers are leaving us because of bugs or usability issues, customer support may have that information. I need all these people as Contributors so that I can get a complete picture of the context in which we're working. And if I know that, then I can make a business case for what we focus on."

"Sparks has been bringing up AI a lot lately," says Sri. "How does DACI help us with that?"

"Well, it's a start," says Irie. "I need enough context to have an opinion about where or whether AI could help. DACI at least provides a guide to who should contribute and how."

Irie adapts a Stakeholder Canvas from her playbook to organize her growing body of stakeholder information. She transfers her stakeholder list and adds two columns: "Their Needs" to capture her stakeholders' goals and priorities, and "My Needs" for what she needs from them. She leaves a few columns blank, anticipating she may need them for more insights as she learns more about Helthex and its people. ■

Figure 1-17. Irie's DACI map for product strategy decisions

Role	People
Driver	Director of product management
Approver	CTO
Contributors	CRO, CFO, director of customer support, VP marketing, VP engineering, director of design, director of data science, product managers, Sparks
Informed	All other employees

Figure 1-18. Irie's Helthex Product Roadmap Stakeholder Canvas

Stakeholder Canvas

Topic Product roadmap decisions — Product, Decision, or Initiative

Driver — Aligns with the Approver on objectives and decision criteria, seeks input from Contributors, and makes the decisions

Irie, director of product management

Objectives

Approver — Sets decision criteria, approves or rejects decisions by the Driver

Name	Role	P/A*	TBD	TBD	TBD	Their Needs	My Needs
Sri	CTO	Ally					

Contributors — Add valuable information, including experts, implementers, and those significantly impacted

Name	Role	P/A*	TBD	TBD	TBD	Their Needs	My Needs
Yacob	Dir. engineering	Ally					
José	Dir. design	Ally					
Sparks	Partnerships	Challenger					
Ella	CRO	Challenger					
Sergey	VP marketing	Observer					
Divya	Dir. data science	?					
Philippe	CFO	?					
Liandri	Dir. cust. support	Supporter					
Liz	CEO	?					
Arianna	Sales	Challenger					
Christina	Product manager	Supporter					
Eitan	Product manager	Supporter					
Min	Product manager	Supporter					

Informed Groups and Individuals — Updated about decisions by the Driver but do not contribute

All other employees

*P/A is Power/Alignment

From the book **Aligned: Stakeholder Management for Product Leaders** (O'Reilly 2024)
For more information, visit www.alignedthebook.com

Aligned

Takeaways

As a product leader, you will likely face challenges in understanding how your organization really works, questions that can't be fully answered by looking at the official company org chart. Like Irie, you may need to develop alternate ways of navigating your organization's structure, influence, culture, and decision-making process:

- Start by understanding whether your organization's structure is primarily Functional, Matrix, Value Stream, Divisional, or some hybrid of these. This will help provide direction about where to look for the most important stakeholders.
- Understand how to identify hidden Power Players by dominant function or by identifying certain individuals as CEO-Whisperers.
- Prioritize your stakeholders using the Power/Alignment Grid, and use the tips in that section to move key players toward alignment.
- Determine your organization's decision-making culture to understand how many people you need to involve in decisions and how much input is expected. Consider our suggestions for influencing decisions to be more Participative.
- Use the DACI model and process to drive alignment on who owns which decisions and who is involved in what way.
- Make a Stakeholder Canvas to keep track of key information about your stakeholders.

After you follow the guidance in this chapter, you may discover, as Irie does, that organizational context is merely the first part of the foundation needed for successful stakeholder management.

Everyone enjoys talking about their own perspective when given the opportunity

Chapter 2

People

Now that you understand your organization a little better, you can start engaging with your prioritized list of key stakeholders. This list will expand over time, but you should connect with the most influential stakeholders now. In this chapter, we will discuss how to get the most out of these initial sessions, preparing you for deeper connections and more durable alignment later on.

As Irie, our product director, meets with her initial stakeholders, we will discuss how to accomplish some key tasks:

- Prepare for and conduct stakeholder discovery interviews.
- Identify and work with different stakeholder decision styles.
- Find time to meet with hard-to-reach stakeholders.

We begin as Irie makes plans to meet with her key stakeholders individually.

2.1 Irie Talks with Stakeholders

Irie stares at her computer, contemplating the list of stakeholders she still needs to meet. It's week four already, and she's barely had a one-on-one meeting with anyone outside Sri's team. She's been trying to schedule time with people, but they always seem too busy to talk with her. Her goal is to get the different functions to talk more with each other, but first she has to actually meet the functional leaders.

She tried to get time with Liandri, director of customer support, but she sent Irie to Zola, a manager on the customer support team. Irie plans to meet with Zola anyway, but will keep trying to get an audience with Liandri in the future.

After introductions and a few pleasantries, Irie shares her thoughts with Zola on the Participative decision model, and how customer support can get involved when product decisions need to be made.

"Maybe we could set up some time to get your team's feedback about what we can do better with the product?" suggests Irie.

"Yeah, I'm sure the team would appreciate that. We're always having to offer customers workarounds for things that don't work in the product. It would be great if you can fix some of those things for us."

"Great! I'll put something on your calendar," says Irie. "Thanks so much for your time."

"Sure thing," says Zola. Then she adds, almost automatically, "Have a great rest of your day!"

"Thanks, you too!" says Irie, smiling, and they both hang up.

Irie is proud of herself for getting Zola's buy-in so quickly. Then her smile slowly turns to a frown. She didn't actually learn anything about Zola's problems in that call, she reflects. All she did was offer her own perspective. Irie knows she tends to get passionate about things when she has new ideas, and she's been working on slowing down and not talking so much. She needs to remember to listen, not just talk.

She decides to write out some questions for her stakeholder interviews to focus the conversation so she doesn't get carried away. *Wait*, thinks Irie, *this sounds just like when I do customer interviews. Maybe I should ask José for some advice. His expertise in user research would be really helpful here.*

Irie still has some time before a meeting with Sergey, but it's cutting it close. She pings José on Slack and discovers he is working from home but is available for a video call. Reminding herself to begin slowly, she asks him how his week is going.

"Going okay," says José. "We're having trouble recruiting participants for one of our research projects. We're looking for people who exercise daily, and apparently there aren't a lot of those out there. Or at least not a lot of them who already use our product."

"That sounds difficult," says Irie. "Let me know if I can help in any way."

"Will do...So what's this about wanting advice on customer interviews? Trying to take my job?" José says cheekily.

"It's only been four weeks but I know enough about you by now to know when you're being sarcastic," says Irie with a smile.

"I think I just need a little advice," continues Irie. "I'm about to do a bunch of stakeholder interviews, and I'm thinking they're probably not that different from customer interviews."

"I never really thought about it that way, but you're totally right," says José. "If you can tell me what you want to accomplish with these interviews, maybe I can help you make a game plan."

Irie thanks José and explains that she wants to understand each stakeholder's role in the company and in product decisions.

"That's simple enough," José says. "Let's talk about how many people to speak with, questions to ask, and how to set up interviews for success." ■

Stakeholder Discovery Interviews

The similarities are strong between stakeholder discovery interviews and customer discovery interviews,* so most of the same techniques can be used. With customer discovery interviews, you're trying to understand how they do their jobs or live their lives so that you can make more informed product decisions. For stakeholder interviews, you're trying to understand how they do their jobs so you can make a plan for how to get alignment and buy-in on product decisions. Either way, you need to learn about their day-to-day experiences and incentives.

Unlike customers, when you talk with stakeholders, you're also trying to get to know them as individuals so that you can form lasting working relationships with them. In addition, it can be helpful to clarify their relationship (current and historic) with the product management function in your organization, so that you can better define your role in that relationship.

Understanding how stakeholders do their jobs, what their incentives are, and how they're measured will help you create a compelling case for your decisions or perspectives when you need to get their buy-in. Knowing what's important to them will help you contextualize their arguments as well.

It's never too late to have a stakeholder discovery interview, whether you just met them or you've known them for years.† You can even re-interview them periodically. The questions may change, but it's always helpful to learn more about someone.

Plus, people appreciate being asked. Most people enjoy talking about their own experience and perspectives when given the opportunity. As Dale Carnegie says in *How to Win Friends and Influence People* (Simon & Schuster, 1936), "You can make more friends in two months by becoming interested in other people than you can in two years by trying to get other people interested in you."

* For more about customer discovery interviews, see our book list at *alignedthebook.com*.

† You don't have to say "I'm doing a stakeholder discovery interview." It's just a conversation where you ask them questions, which can be done very casually. And you can ask questions over time in different conversations; it doesn't have to be all at once.

How many stakeholders?

From Chapter 1, you should now have a list of key stakeholders to speak with, and talking with each Power Player individually will be critical. But if you work at a larger organization, once you get past the Power Players, you may just have a list of departments and not know who in those departments to speak with. Since you can't reasonably speak with everyone in a department, try to start with the person as high in the department as possible.

For example, if your marketing department is five people, speak with the head of marketing. If your marketing department is 500 people, try to find the leader of the team within the marketing department who is most closely aligned with your Product Team. Speaking with the person at or near the top is useful because they can provide strategic context and help you understand who on their team to connect with about different topics.

We recommend meeting one-on-one as the most efficient and effective way to learn what you need to know about an individual. Speaking with one person at a time allows you to ask follow-up questions and go deeper than is possible in group meetings. Talking with stakeholders individually may also make honesty easier: they may not want to speak openly in front of a crowd.*

Questions to ask

A stakeholder discovery interview is about learning. It's not about alignment (yet), or pitching your idea, or establishing rank. It's about listening more than talking so that you will have a basis for having productive conversations about specific decisions or issues later.

First, you should set an objective for the conversation. Are you just meeting them and want to understand how their department operates? Are you starting a new project together and want to know more about how their department incentivizes performance? Have you known them for years

* Stakeholder discovery interviews are best done one-on-one, but sometimes it's okay to talk with stakeholders in pairs, especially if your stakeholder requests a third person to join the meeting. You can always regroup with your stakeholder another time to get to know them individually.

but are now starting to work together in a new way and want to understand their perspective?

With your objectives in mind, make a list of questions designed to help you get the right information. A few tips to think about when creating your questions:

- Ask open-ended questions using phrases like "tell me about" or "what's challenging about." Their own words will tell you more than yes/no or multiple-choice options.
- Start with high-level questions to develop context and put them at ease. Save more specific questions for later, so you don't bias their answers.
- Plan on one question for every five minutes you have with them, knowing you may not get to them all. So, for a 30-minute meeting, pick the six most important questions, and expect to get to three or four of them.
- Ask follow-up questions if an answer is vague, like "Tell me more about that" or "Can you give me an example?"

On the next page are some stakeholder discovery interview questions we like to use (Figure 2-1).*

* You can find a more comprehensive list of questions on our website: *alignedthebook.com*.

Figure 2-1. Sample stakeholder discovery interview questions

Question	Why to ask
Tell me about your role	This will get them to explain in their own words what they actually do rather than simply what their title or job function suggests that they might do.
What do you like best about your job?	This fun question allows them to express what they value and gives you clues as to what proposals may appeal to them.
What does success look like for you and your team?	This question allows them to explain what they view as most important to them over time and provides context to all of the rest of the interview. They may or may not mention objectives or metrics here.
What's challenging about meeting your goals (or metrics)?	This is where you learn about the "pain" or "problems" that motivate them. Keep it open-ended and let them talk. Don't limit it to areas your product serves or your team can obviously help with.
What are you doing to try to meet your challenges?	You can learn a lot about how they think and operate, who they work with, and how, by letting them talk about this. It also helps you understand how anything you might propose will be evaluated.
What's top-of-mind for you right now?	This question helps you understand what they are focused on right now, and might help you understand how they view you, your product, and your team.
How do you or your team usually work with my team (if you do)?	This question helps you understand their perceptions of your function and your particular team. It's also a way for you to invite critical feedback on yourself indirectly. It can be easier for them to provide you with negative feedback by implying it's not just you.
If you could wave a magic wand and change anything in this organization, what would you change?	The classic wish-fulfillment question gives them space to offer ideas or share challenges that haven't yet come up.
What should I have asked that I didn't?	The ultimate open-ended question, it helps to save this one for last to give them space to offer things you didn't think to ask about.
Who else should I be talking to?	This helps you understand if they need someone else's input before buying into anything you might propose. It also helps you fill out your Stakeholder Canvas.

Use probing questions (Figure 2-2) as follow-ups to an initial answer to get more detail. You can use multiple probes in sequence to learn all you can about a topic.

Figure 2-2. Helpful probing questions

Type of probe	Sample questions
Clarify the answer	"Can you give me an example?" "What do you mean by 'confusing'?" (Or 'annoying,' 'difficult,' 'complicated,' 'problematic'...) "It sounds like you were saying..."
Feelings, motivations	"What was the goal?" "How did you decide to do it that way?" "What made that part so important?"
Get more details	"Tell me more about that." "Echo" what they just said to encourage them to continue. "How so?" "Can you give me an example?" "Tell me about the last time that happened." "Anything else?" "Do you always do it that way, or do you sometimes do it differently?"

We strongly recommend asking open-ended questions that invite the other person to share as much as they are willing. However, after establishing some shared context, you may have developed some preliminary conclusions about a particular stakeholder and their views. At that point, you can attempt to validate your thinking by asking leading questions. If, for example, they work for the customer support team, you may say something like, "I imagine you are measured on things like time to resolution and call length. Is that right or are other things more important?"

Tips for stakeholder discovery interviews

Many of these suggestions will sound familiar if you've led or participated in user research interviews. The goal with each of these tips is to make it as easy as possible for your stakeholder to share what's on their mind.

Create a comfortable environment

Consider noise level, confidentiality, and possible distractions when choosing the setting for your conversation.

- **Make it informal:** A small conference room is fine, but the informality of a company break room, lunch room, or even a nearby cafe may help put your stakeholder at ease.
- **Keep it confidential:** If you are in a public space (even on video or by phone), maximize confidentiality by sitting away from others and keeping your voice low.
- **Minimize interruptions:** Silence notifications on your devices, including your computer itself if you are meeting via video.
- **Allocate enough time:** Start with 30 minutes for your first few, but if you often find yourself with more questions at the end, consider making future sessions an hour or booking follow-up time.

Find your way of taking notes

Some people are uncomfortable being recorded, so for these personal one-to-one meetings, we don't recommend using your phone or video conferencing platform to record.* Instead, take notes.

- **Use pen and paper:** This informal method encourages stakeholders to relax and be more forthcoming.
- **Type notes in real time:** This saves you from having to transcribe your notes later, but be aware that it may feel more official and may cause you to look away more.

---

* This is evolving as more and more people use automated tools to record, transcribe, and summarize meetings. If you decide to use any form of recording, ask permission before turning it on, and be respectful if some or all of your stakeholders decline. If you have permission to record and your stakeholder is hesitating to talk about a sensitive topic, offer to turn off the recording. Being gracious about this is a quick opportunity to earn respect.

- **Don't invite a notetaker:** Asking someone else to take notes may be a good tactic for customer interviews, but it violates the intimacy of the one-on-one relationship you're trying to build with your stakeholder.
- **Take notes after:** Immediately after the interview, take a few minutes to write down key points from the conversation when your recall is best.

Helen Saunders, VP of product, says, "I like to say up front that I'll be taking a few notes to help me capture their inputs. If we are on video, I say that if I'm looking down or to the side then I'm writing in my notepad, so they don't think I've lost interest."

Follow their lead

Letting your stakeholder lead the conversation helps you understand what's most important to them (as long as they're not going on too much of a tangent).

- **It's about them:** Ask a question and then stop talking. Listen and give them time to explain. Ask clarifying questions that help you understand, while interrupting them as little as possible.
- **Make it a conversation:** Follow the natural flow of conversation and insert your questions where they seem to make sense, rather than reading them in order like a script.
- **Choose what to ask:** Aside from a few "must ask" questions, you can have different conversations with different people. Going deep on individualized topics of interest will show you are listening and enhance your understanding of each person. You can always follow up later.

Use active listening

The purpose of this interview is for you to learn, so it's important to do more listening than talking. When people feel heard, you learn more because they are more likely to be honest.

- **Don't talk about you:** Don't offer your own opinions, judgments, or decisions unless asked. You are there to understand their point of view fully, not to judge or persuade.
- **Summarize:** Reflect back what you think you've heard from them and ask if you got it right. This often leads to important clarifications of points you may have missed or misunderstood.
- **Make connections:** Connect with what was said earlier, such as, "Earlier you said...Can you tell me more about that?" or "Just like you were saying earlier."
- **Embrace the silence:** Let them think and respond, don't answer your own question, and let them talk before using your own clarifying examples.*

Liz Lehtonen, product director, shares that "the first meeting is proving how well I listen."

* This is the hardest one for us!

PRO TIPS

Typical Metrics by Department

You may begin to understand stakeholders' motivations by recognizing typical incentives by department (Figure 2-3). We recommend confirming these with your stakeholders, but these are a good starting point.

Figure 2-3. Typical metrics by department

Sales

Sales cycle length, win/conversion rate, percentage of team achieving quotas, renewal rate, contract value, retention rate, churn rate

Customer support

Labor cost, contacts per order, first contact resolution rate, resolution cost, interactions per ticket, customer satisfaction score, net promoter score, ticket volume, time to resolution

Finance

Revenue, working capital, accounts payable turnover, return on equity, gross profit margin, earnings per share, operating expense, EBITDA

Field operations

Units per hour, labor cost, on time rate, safety incidents, efficiency, order backlog, damage rate, response time, inventory accuracy, throughput

People/human resources

Employee satisfaction (like eNPS), offer acceptance rate, retention rate, cost per hire, diversity statistics, benefits adoption, safety incidents, revenue per employee, training completion rate

Professional services

Billable hours, utilization rate, project profitability, project margin, client retention rate, services revenue, forecast demand total

Customer success

Renewal rate, churn rate, customer retention cost, customer lifetime value, net promoter score, customer health score

Engineering

Team velocity, system uptime, incident rate, deployment frequency, mean time between failures, mean time to recover, test coverage, change failure rate

Marketing

Conversion rate, click-through rate, bounce rate, cost per lead, customer acquisition cost, customer lifetime value, brand awareness, qualified leads, impression share, net promoter score

Design

Product adoption rate, average session length, abandonment rate, task success rate, time on task, monthly active users, user retention rate, net promoter score, system usability score

Irie Conducts a Stakeholder Interview

After speaking with José, Irie feels she has a much better plan for how to talk with her stakeholders. She's worked out a short list of questions with José that she plans to use moving forward.

Figure 2-4. Irie's list of stakeholder questions

1	A
2	What does your team do?
3	What does success look like for you and your team? How do you measure success?
4	How have you historically worked with product management?
5	How do you feel about AI going into the product?
6	Who else should I be talking to?

Looking at her list of questions, she adds a list of follow-ups at the bottom, like "Tell me more about that" and "Can you give me an example?" to remind her to probe for more details. She finishes just in time for her interview with Sergey.

Irie arrives at the meeting room to see Sergey sitting with a colleague she hasn't met yet. The pair are deep in conversation. Sergey has his arms crossed in front of him as he speaks in a tense tone. He stops speaking and drops his hands to the table when Irie enters.

"You must be Irie," says Sergey in a friendlier tone. "I'd like you to meet Alex. They handle market research and analyst relations on my team. I brought Alex because I think that they will be the one working most closely with product management."

Irie makes a mental note of Sergey's use of "they" to refer to Alex and resolves to use that pronoun for Alex to show she is listening and being respectful. "Great to meet you, Alex," says Irie, reaching out to shake their hand as she sits.

"You as well," says Alex. "I've heard a lot about you. I'm excited to see how we can work together."

"Me too," says Irie, taking out her notepad. "Mind if I take a few notes?"

"Of course," says Alex. Sergey nods as well, with a friendly smile.

"Alex, can you tell me a bit more about what you do?"

"Sure," begins Alex. "I do market research, learning about what thought leaders, analysts, and industry panels are saying so I can better understand our market. I also produce reports for Helthex leadership."

Sergey adds, "In addition to Alex's work, the marketing team handles lead generation for corporate sales and customer acquisition for direct-to-consumer sales. We also handle branding and public relations."

"That's helpful, thanks," says Irie. "I imagine that Alex's research supports a lot of the work you do in the other parts of the team?"

"Yes," says Sergey, "their work has given us invaluable input into our marketing strategy."

Irie asks a few follow-up questions, and then moves on to the next topic on her list. "What does success look like for your team?" Irie asks.

"Our biggest challenge right now is acquiring new customers," admits Sergey, "so conversion rate is something we're focused on. The big corporate deals bring us a lot of users at once, but the cost of acquisition is higher after sales commission is factored in. For direct customer sales, we're working on creating more awareness."

Irie notices that Alex hasn't said much, so she poses the next question to them directly. "Alex, what have you discovered so far in your research about customer conversion?"

Alex explains, "I've been interviewing customers to learn what motivates them. I've only had time for half a dozen interviews so far, but a lot of them say that the main reason they use our app is the customized advice."

Irie nods encouragingly, and Alex continues. "Some complain that our advice is not specific enough to them, though, and they want it even more customized and personalized. One person even said that they expected us to find correlations in their own data and alert them to dangers or small lifestyle changes they could make that would have a positive effect on their health."

"That makes sense," Irie replies. "Like, 'We've noticed that you sleep better if your last cup of coffee is before 2:00 pm.'"

"Exactly," Alex says while Sergey nods.

Sergey chimes in then. "We've been thinking about how our basic recommendations model could get to this level of detail with all of our users, but that doesn't seem scalable. I guess that is where we could use your help."

"This makes me think of Sparks's idea of using AI," Irie says. She notices an angry expression appear on Sergey's face and he crosses his arms across his chest. Assuming he is reacting to her mention of Sparks, she says, "You've probably gotten an earful about this as well. But maybe that's something the technology team can use to get to the level of detailed recommendations you're talking about."

Sergey looks down at the table, arms still crossed, avoiding eye contact. "I think that is a very dangerous idea," he says. "Think of the liability if someone is harmed by following our AI's recommendation. And the people who would be out of a job!"

"Yes, I'm sure that's a concern," says Irie thoughtfully. "Perhaps we could..."

Before she can finish, Sergey cuts her off. "I believe it is immoral, asking machines to make decisions or substitute their judgment for advice from our panel of trained medical experts."

"That's understandable," Irie says, remaining calm.

Before she can say more, though, Sergey closes his laptop with an audible thump. "I just noticed the time," says Sergey, who has not consulted his watch. "Sorry, I have to run, but Alex can answer any other questions you have." Sergey picks up his laptop and leaves the room.

Irie looks quizzically at Alex.

"Sergey gets like that sometimes," says Alex. "He can be quite...passionate about things."

"I can see that," says Irie. "Any idea what's going on?"

After a moment, Alex decides to share what they know. "I don't know if this is really what is bugging Sergey, but he expressed some frustration this morning about his brother losing his job as a writer because his company decided to use AI instead for content. I think he's a bit sensitive about this right now."

"I wish I had known—I would have brought up the AI topic in a more delicate way!" says Irie, reflecting on the conversation. "I think my problem is that I jumped right to business in this meeting. I was so set on my agenda, on my questions, I forgot to ask about what was on Sergey's mind, or on yours. I didn't even ask 'How are you?'" Then after a moment, she adds, "No wonder he jumped right to how you two could contribute."

"It's okay," says Alex. "I do that too. I forget to do the small talk at the beginning of the conversation. It just doesn't come naturally to me, so I have to constantly remind myself to be more personable." Alex pauses. "Which is funny, because in my job I really enjoy figuring out what people are thinking and how they behave. I have to do that in order to understand customers' motivations and purchasing habits. I guess I forget to do it in my own interactions at work."

"Same here," says Irie. "I talk with a lot of people in my job too, and I have to remember that they're people and not just sources of information."

"I know, right?" says Alex. "Sometimes I need a break with a good spreadsheet after a few interviews," they add with a laugh.

Irie laughs along with Alex and admits she has the "spreadsheet gene" as well. "There's just something comforting about everything adding up."

"Exactly!" Alex says, making a gesture toward their laptop.

Irie looks at her watch and shares that the meeting time is now over for real. "Thanks for giving me that insight into Sergey's feelings about AI. I really appreciate it."

"No problem," Alex replies, offering a fist for Irie to bump. "Spreadsheet geeks gotta stick together." ■

2.2 Irie Adds Decision-Making Styles

A few days later, Irie is still having trouble getting in touch with Sparks. She has emailed him, Slacked him, put a meeting on his calendar (which he declined—or maybe his admin, Pria, declined it) and she even tried calling the phone number in his email signature. No response.

Irie is also having trouble getting in touch with most of the other stakeholders on her list, except for Ella, who she's talking to in a few days, so she decides to move on to other things for now. She has a one-on-one coming up with Christina, a PM on her team, so she decides she will see if Christina has any more info on Sparks.

Irie joins the call with Christina and realizes that her background looks familiar. "Are you in the office?" she asks.

"I wasn't planning on coming in, but I had a chance to meet with Liandri because she's in town this week from South Africa, so I jumped on it," she explains. Irie wonders why Christina is able to get a direct meeting with Liandri, while *she* gets delegated to someone on Liandri's team.

"Well, if we're both here, we may as well meet in person," says Irie. She proposes they meet in the large kitchen near her desk. She grabs her laptop and walks to a large, well-lit room with colorful furniture, two elaborate automated coffee machines, an array of snacks, a cold-brew machine, and four beer taps. Someone is using one of the coffee machines, but since it's past lunchtime, the room is mostly deserted. She spots Christina by herself on a small couch, and approaches.

"Thanks for meeting in person," Irie says, and chooses a large ottoman across from the couch. "I always find that speaking face-to-face leads to better communication and understanding." Irie puts her laptop down on a small side table. "You came in to meet with Liandri, so you must appreciate that."

"Oh yes," says Christina. "Any time I can get face time with an exec, it's hugely valuable."

"I can understand that," says Irie, lamenting her own inability to get stakeholders to talk with her virtually or in person.

"Sparks is even harder to get in touch with," says Christina. "Liz is checking in occasionally, but she's still on medical leave, so her time is precious these days." Christina explains that she checks her priorities with these two executives regularly.

When Irie asks why, she replies that the founders are ultimately in charge. "They built the company, and I find it's best to follow their lead. With Liz mostly out, I'm always asking myself, 'What would Liz do?'"

"But I thought our DACI process was supposed to fix that?" asks Irie.

"It helps with product decisions, for sure," says Christina. "It's much easier to make decisions now that we are deliberately choosing a decision maker for each one." Christina pauses. "But Sparks still comes with his requests, and that messes up our roadmap."

"What has he asked for recently?" asks Irie.

"Since my focus is on the analytics side of things, Sparks usually comes to me for integrations with other apps and devices that get us more data. Many of those integrations are helpful, but some are just distracting. It doesn't seem like he's prioritizing which data is most important and what we're going to use it for."

"If Sparks wasn't asking for things," says Irie, "how would you make the decision of what to work on next?"

"I've been asking for a while to collect data on how our users actually use the app day to day," begins Christina, "but that never makes it onto the roadmap. If I knew more about how users were actually using the app, I could figure out what the next logical data source should be. I also get some information from sitting in user interview sessions, but that's mostly qualitative. I haven't had any luck using data to convince Sparks of anything. He just doesn't seem to think like I do."

"That makes sense," says Irie. "We all make decisions differently. I value making decisions quickly, too, even if we don't have all the information we need, but I like to have at least some data."

"Sometimes I can get into a bit of an analysis paralysis," Christina admits. "But Sparks seems to make decisions with no data at all. Like he's happier with a random decision than no decision. Doesn't make sense to me."

"And Liz?" asks Irie.

"Liz is the big visionary, always talking about her ideas for the future. It's motivating, and it's why a lot of us are here. But she usually leaves the detailed decision making to other people."

After Irie and Christina finish their conversation, Irie remembers that she has a section on decision styles in her playbook, just after the slide on organization decision culture. She brings it up on her laptop to review it. It starts by explaining that organizational decision-making culture helps you understand how groups make decisions. But awareness of individual decision-making styles helps you understand how individuals make decisions that are assigned to them. Understanding individual styles is most helpful in Directive and Participative approaches, where there is a single decision maker who makes the ultimate call. ■

Individual Decision-Making Styles

In Chapter 1, we describe how organizations make decisions. Individuals have their unique approaches to decision making as well, and understanding someone's individual decision-making style helps you approach them in the right way with the right information.

The DiSC* personality types provide a simple way of understanding how different people typically think and act. We've summarized how the DiSC types apply to decision making in Figure 2-5 for easy reference.

Figure 2-5. Individual decision-making styles

Dominance

Motivator
Wants a quick answer to act on

Approach
Get an answer quickly

Preferences
Minimal discussion

Characteristics
Intuitive, strong-willed, forceful, confident, direct, rash

Benefits
Efficient decision making

Risks
Fast decisions are not always the best ones

How to work with them
Tell them that including more perspectives doesn't have to slow them down. Remind them that even the best ideas should be tested before being adopted, which will avoid future slowdowns.

Influence

Motivator
Wants to do the right thing

Approach
Begins with a big idea, gets people motivated

Preferences
Comfortable with "what ifs," enjoys debate

Characteristics
Conceptual, outgoing, enthusiastic, optimistic, persuasive

Benefits
Motivation, team alignment

Risks
Can be impractical, lack of follow-through

How to work with them
Frame your ideas in terms of the benefits to the big picture. Remind them that ideas need to be practical and that data can help determine what will be successful.

---

* DiSC was developed by William Marston in the 1920s. Personality assessment tools were developed later and have become common tools for understanding the links between personality and behavior. Fun fact: Marston also developed the polygraph machine. See *https://discinsights.com/pages/william-marston-disc*.

Steadiness

Motivator
Wants what's best for everyone

Approach
Seeks input from wide range of people

Preferences
Wants everyone to agree before moving on

Characteristics
Empathetic, agreeable, flexible

Benefits
Alignment is almost guaranteed

Risks
Suboptimal decision due to a mediocre compromise

How to work with them
Show the benefits to everyone involved. Remind them that it's impossible to please everyone and sometimes it's better to make a decision quickly than to make the perfect decision.

Conscientiousness

Motivator
Wants an optimal answer

Approach
Begin with facts, develop logical arguments

Preferences
A lot of data, considers many options

Characteristics
Analytical, precise, systematic

Benefits
Decisions are often optimal

Risks
Slow to make decisions, can get stuck in "analysis paralysis"

How to work with them
Provide data to influence them, and be patient as they work through the logic. Remind them that perfect knowledge is not possible and it's best to make choices that drive toward the overall vision.

How to determine someone's individual decision-making style

Short of an elaborate personality test, we've found that people are generally open to discussing their approach to decision making. The simplest way to learn about their typical style is to ask them. To avoid generalities, ask them how they think *a specific decision* should be made or was recently made. Their answers will help you determine their primary style (Figure 2-6).

Once you have a good sense for how your individual stakeholders make decisions, you will be better equipped to deal with challenges that come up in the decision-making process. Understanding someone's decision-making style, especially if it's different from your own, can help you reframe your narrative to appeal to the way that person thinks. For example, if you make decisions in the Conscientiousness style, you might attempt to influence your stakeholder with all the data you can find. But if your stakeholder falls into the Steadiness bucket, a better approach might be to explain why your proposal benefits all involved parties.

Figure 2-6. Questions to determine individual decision-making style

Question	Dominance example	Influence example	Steadiness example	Conscientiousness example
What result are you seeking with this decision?	"I just want to make the decision quickly so we can move on."	"I want to do what's best for the company."	"I want everyone to be okay with the direction we choose."	"I want to make sure we've considered all our options."
How should we go about deciding?	"We should make the best decision we can, given what we know now."	"We should make a decision that aligns with our vision."	"We should get everyone's opinion before we decide."	"We should gather all the data before we decide."
Who should we involve in making the decision?	"We only need a few key people."	"I hope that everyone will want to be involved."	"We'll need representatives from every department."	"The data should speak for itself."
How do you feel about this decision?	"Let's do it and see how it goes."	"I'm really excited about the direction we're heading."	"I'm glad we were all able to agree."	"The data points to this being the right decision, so I'm optimistic."
Did you expect a challenge from anyone on the decision or the approach?	"I expected [Steadiness person] to worry too much about how people will feel."	"I expected [Conscientiousness person] to get into too much detail."	"I expected [Dominance person] to try to rush things without getting everyone's opinion."	"I expected [Influence person] to overlook important realities while pursuing their vision."

Irie Adds Individual Decision-Making Styles

Irie adds a column for DiSC to her Stakeholder Canvas and begins categorizing her stakeholders based on what she knows so far. José is thorough and methodical, she thinks—classic Conscientiousness. She writes the same for Sergey. Yacob seems focused on what's best for his team; she writes Steadiness for him. Sparks is clearly a Dominant type. She puts a "?" in for people she hasn't gotten to know well enough yet.

Figure 2-7. Irie's updated Stakeholder Canvas with DiSC

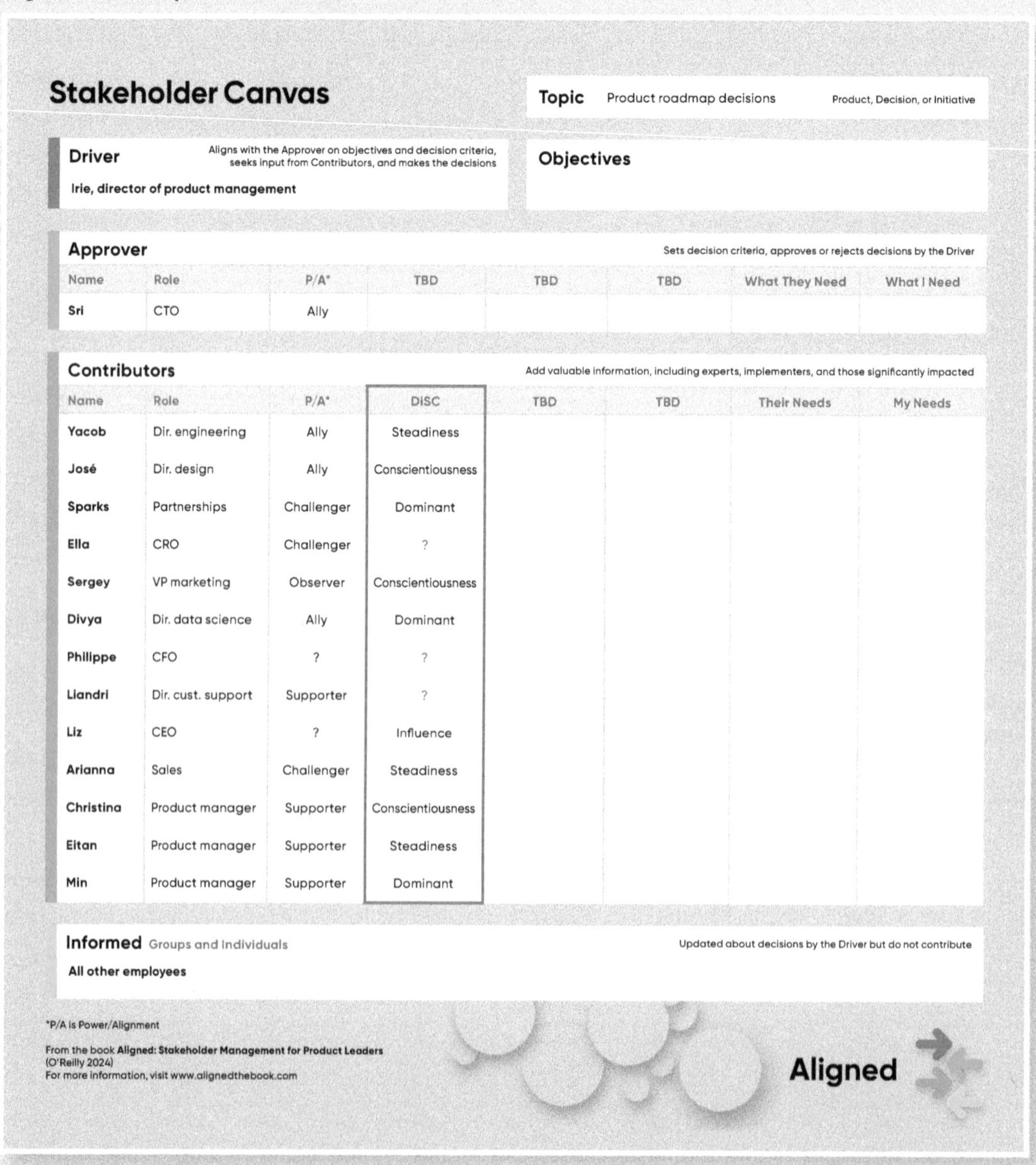

Stakeholder Canvas

Topic Product roadmap decisions — Product, Decision, or Initiative

Driver — Aligns with the Approver on objectives and decision criteria, seeks input from Contributors, and makes the decisions

Irie, director of product management

Objectives

Approver — Sets decision criteria, approves or rejects decisions by the Driver

Name	Role	P/A*	TBD	TBD	TBD	What They Need	What I Need
Sri	CTO	Ally					

Contributors — Add valuable information, including experts, implementers, and those significantly impacted

Name	Role	P/A*	DiSC	TBD	TBD	Their Needs	My Needs
Yacob	Dir. engineering	Ally	Steadiness				
José	Dir. design	Ally	Conscientiousness				
Sparks	Partnerships	Challenger	Dominant				
Ella	CRO	Challenger	?				
Sergey	VP marketing	Observer	Conscientiousness				
Divya	Dir. data science	Ally	Dominant				
Philippe	CFO	?	?				
Liandri	Dir. cust. support	Supporter	?				
Liz	CEO	?	Influence				
Arianna	Sales	Challenger	Steadiness				
Christina	Product manager	Supporter	Conscientiousness				
Eitan	Product manager	Supporter	Steadiness				
Min	Product manager	Supporter	Dominant				

Informed Groups and Individuals — Updated about decisions by the Driver but do not contribute

All other employees

*P/A is Power/Alignment

From the book **Aligned: Stakeholder Management for Product Leaders** (O'Reilly 2024)
For more information, visit www.alignedthebook.com

Aligned

Irie wonders idly how she would classify herself. Given what she told Alex about her "spreadsheet gene," she is tempted to pick Conscientiousness. But she knows she can be decisive and directive when it's called for, so there is at least some Dominance there. She also really cares about her team, and she is spending a lot of energy trying to get to know all the key players, so there is Steadiness in her as well. *What about Influence? Is she a big-picture person, too? It's hard to be objective about oneself,* she reflects, *but maybe product managers have to have a little of everything.*

After she updates her Stakeholder Canvas, Irie prepares to meet Liz for the first time. She is excited and nervous all at once. She knows that a lot of people joined the company because of Liz's charismatic leadership style (Influence type, she thinks) and compelling vision of providing "personalized medical advice for anyone, anytime, anywhere." Even though Liz was on medical leave during Irie's interview process, Liz's vision was still a compelling reason to join.

She logs into the meeting and ponders why Liz seems to be easier to schedule a meeting with than Sparks, even though she's still technically on leave. Liz joins the meeting shortly after Irie but her video is just a black box with a bright red "L" in a circle in the middle. Irie wonders how she's feeling.

"Hi Irie," says Liz, "it's great to meet you." From behind the blank screen, Liz's voice is slow and soft and steady. "Oh wait, my camera's off, hold on a minute."

Liz turns on her camera to reveal a smiling woman against a blurry background through which Irie can just make out a bookshelf and a space heater. Liz is wearing a fuzzy blue sweater and a giant gray newsboy cap. Wisps of blond hair are coming out around the sides of the wool hat.

"Hi Liz," says Irie, "I'm so happy to finally be meeting you."

"Me too," says Liz. "Sparks has told me great things. He said you started contributing even on day one."

Irie remembers back to her first meeting at Helthex and how she got off on the wrong foot with Sparks. She's not sure if he was being sarcastic when he said that to Liz. "Thanks," says Irie. "I hope I can live up to the stories."

"Sri has also spoken very highly of you," continues Liz. Irie thinks that sounds more likely. "When I was in the office more, I always took new hires out to lunch on their first day, but that's a bit harder right now, for obvious reasons." Liz smiles. Irie smiles too, relaxing a bit but still somewhat uncomfortable, not knowing how to react to Liz making jokes about her illness. "So I've been trying to at least meet people remotely before too much time passes."

They chat for a bit longer and then Liz asks Irie if she has any questions for her. Irie remembers her stakeholder interview questions list, but she's not sure if those questions make sense in this context. She thinks about decision-making styles and how Christina said Liz is in the Influence category. She decides to start with a different question. "Part of the reason I joined Helthex is because of the company vision, our commitment to 'personalized medical advice for anyone, anytime, anywhere.' Since I'm starting to think about how the product will help achieve that vision, is there anything else you can share about where you see the product going?"

"That's a great question," says Liz. "Let me start with a little background about why I founded the company. My parents both died young. My mother from a form of cancer that slowly spread throughout her body. There were signs, but by the time she went to see a doctor it was too late. My father was a smoker and he had heart troubles. He was too depressed after my mother died to see anyone. Things got worse and he was gone, too, within a year."

Though she recounts this story with practiced smoothness, her eyes redden as she speaks and she pauses before continuing. "I started the company because I wanted to create a product that helped answer people's questions, because people often can't or won't see a doctor, the internet is too impersonal, and navigating illnesses alone is not easy. The company has to make money, of course, but I want to focus on positively impacting people's lives. I want you to make sure we're helping people when they need it most."

"I'm so glad you want us to focus on customers," says Irie. Then she thinks of another question. "With such a personal mission, how did you end up with Sparks as a cofounder?"

"I've known him for a long time," says Liz, smiling. "He's a good businessman and a brilliant technician. I needed him to help me turn my ideas into reality."

Before Irie has a chance to ask any follow-up questions, Liz's phone rings. She looks at it and says, "I'm sorry, I have to take this. But I'm so glad I got to meet you."

"Me too," says Irie.

"I'm only working a few hours a week right now," Liz says, "but when you have a product roadmap put together, I'd love to see it. Reach out to Pria to set up time." Liz mutes and starts speaking into her phone. She smiles and waves at the camera and then disconnects. ■

2.3 Irie Approaches Busy Stakeholders

The day after her conversation with Liz, Irie finally has a chance to talk with Ella, the CRO.

Ella dials into the call from the back seat of a rideshare, apologizing for not meeting earlier. "It was a bit frantic trying to close a few key deals right up until the last minute," she explains, "and I am still catching up."

Following Ella's lead, Irie asks about their main sources of revenue. She learns that most of their users are consumers who sign up directly. "But we have been trying to grow our corporate business this year," she adds. As Ella elaborates, Irie gathers that the corporate business is not going well. Ella says that they lack many features large corporations require. "We lose most of our large RFPs and we have to discount heavily to win the rest."

This confirms what Irie learned from Eitan, that the sales team asks for a lot of features for big prospects. Ella says she would like to see the technology team focus more on these missing corporate features, but she's not sure it will be enough. "We've asked for so many features, and we just don't seem to be making progress. I know you just got here, so it's not your fault. But it feels like we will never make them all happy."

Irie works through her other questions. Ella is very open to ways they can work together to improve decision making. "I'd like to see more people involved in the decision process," Ella shares, "not just one or two people. Lots of people around the company have great insight to offer." This reminds Irie of the Steadiness decision style, something Ella apparently shares with Eitan. She also thinks that she should find a way to help the product management team manage all the incoming requests from around the organization.

Irie thanks Ella and then asks if she has any advice for connecting with the other executives on her list. "I'd love to get more input," she explains, "but you, Liz, and Sergey were the only ones who got back to me."

Ella laughs sardonically. "Yeah, our bunch isn't always great about collaborating with people from other departments," she says. That makes sense to Irie based on the Functional org structure. "I admire your persistence. Let me give you a few tips from my time as a quota-carrying salesperson."

Ella shares how she connects with busy executives who don't easily make time for people. Irie takes notes and, laughing at some of them, decides she should add a new section to her playbook. ■

The "Unmeeting"

The higher up your stakeholder is in the organization, the more difficult it can be to get their time and attention. For the most elusive stakeholders, consider an "Unmeeting."

An Unmeeting is an informal (often short) conversation you can fall back on when you can't arrange a formal meeting. These generally take place asynchronously or while the stakeholder in question is doing something else, like walking between meetings, eating lunch, or riding in a taxi. You can even get time with a busy stakeholder by joining another activity they're involved in like a company sports league or a volunteering event (assuming they have time for something like that).

Aside from the few minutes you're able to spend with the stakeholder, a major benefit of the Unmeeting is that they will often agree to set up something more formal soon thereafter, once they get a taste of the topic you want to discuss. Plus, if you're nice, they may give you the secret to getting a formal meeting onto their calendar, like "Talk with my admin, Carmen. She takes care of my calendar. Tell her I said she should make time for this meeting on Monday afternoon."

How to arrange an Unmeeting

Even though Unmeetings can be somewhat spontaneous, they benefit from advance planning. There are a few ways to arrange an Unmeeting.

Seek a helper

The best way to arrange an Unmeeting is to get to know the busy stakeholder's administrative assistant ("admin") or executive assistant ("EA") if they have one. This person is usually a hidden Power Player because they control the calendars of the more obvious Power Players. Building a good relationship with them is key to unlocking time with your busy stakeholder and getting the inside scoop on their schedule. It may sound obvious, but treating admins like human beings and colleagues, rather than functionaries, is a surprisingly underutilized approach.

Another way to reach a busy stakeholder directly is through the chain of command. If your local (or organizational) culture is especially respectful of rank and protocol, and you need to connect with someone higher up, seek an introduction via your functional leader of equivalent rank. If you want to speak to the CFO, for example, ask the C-level exec that your group reports to to reach out on your behalf.

Learn about their habits

If your busy stakeholder doesn't have an admin, you can try to learn more about their habits through observation or asking your peers. For example, if you're in a physical office together, maybe you've noticed that they're always in the office already when you arrive at 9 a.m. Start arriving earlier to see if you can find out more about their morning routine.

Melissa prefers to work at the early end of the day, and at one job she would regularly show up at 8 a.m. A high-ranking operations stakeholder also showed up at that time, and Melissa noticed that he would spend 15 minutes every morning making pour-over coffee in the tiny kitchen near her desk. So she started making her tea in the office instead of at home, to have the opportunity to chat with him before they both started their day.

Meet them where they are

You can bring your laptop and sit in common areas, like the lunch room, to see if maybe your busy stakeholder will walk by. If you see them in the hall, offer to walk with them to their next meeting and have a quick chat along the way. Or if you see someone you want to talk to in a meeting room, walk by at the end of the hour or half hour. Or, if possible, look up the meeting room calendar to see when the meeting ends. Then "happen to walk by" at just the right time.

Joshua Herzig-Marx, startup founder turned advisor, tells a story about a colleague he thought was dodging him. "I couldn't get time with him. He canceled any meeting I put on his calendar. It was weird."

"So one day I decided to up the 'weird' and offered to walk him to the train station. I didn't pretend that I was going that way anyway. In fact, I was pretty clear. I said: 'We need to talk about roadmap stuff, it's been really hard to find time with you. Please tell me if you'd rather I not walk with you to the station.'" Joshua's directness caught this stakeholder off guard enough that he agreed. Even better, it set a direct and pragmatic tone for the conversation.

Chat asynchronously

In a virtual environment, sometimes you can get an Unmeeting asynchronously. Plain old emails seem to be going out of favor these days, but some people respond well to them. If you have a messaging culture, use Slack or Teams to reach out. Sometimes a few quick lines will convince them to set up an impromptu call or video chat for a few minutes, which can be a thousand times more effective than asking to schedule a meeting.

Personal apps like WhatsApp (especially in Europe) or WeChat (in China) are great for a quick connection, but be careful with this one. You need a preexisting trusting relationship to utilize this channel, otherwise your stakeholder could be turned off by the presumed closeness of this approach.

Melissa once messaged the CTO of her 15,000-person organization on Slack because she had a potential candidate for a high-level role. He redirected her to the person recruiting for the role, but he did take a few minutes to chat about the need for more women in high-level roles at the company.

Questions to ask

Use your stakeholder interview questions in Unmeetings as well, but adjust your approach to the setting. If the meeting is rushed, ask one question at a time and keep it short. If you are chatting over coffee and you have more time, ask an open-ended question like "What are you working on right now?" or "Can you tell me more about what your team does?" In a more relaxed setting, you can also ask questions that might take some time to answer, like "I'm working on X, and I'm running into problems. What's your perspective?"

Some Unmeetings are much shorter, like a walk down the hallway between meetings, an elevator ride, or a minute at the beginning of a virtual meeting where you and your stakeholder are the only ones who have joined so far. In these cases it's best to ask a short, simple question, even a yes or no question. Some examples: "I'm working on X. Is that something you'd like to have input on?" or "I've decided to do A instead of B. I just want to make sure that doesn't conflict with your plans."

For any Unmeeting, it's always wise to ask for a "real" scheduled follow-up meeting. Something like: "I'd like to talk more about this, if that works for you. Can I follow up with your admin to get time on your calendar?" They may not have time for a follow-up, but it doesn't hurt to ask.

Unmeeting Cautions

The main benefit of an Unmeeting is that you get time with a hard-to-reach stakeholder. But there are some downsides:

- With limited time you can't ask all the questions you could in a longer discussion.
- Your stakeholder is likely distracted or multitasking and may not answer your questions as thoughtfully as you'd like.
- Your stakeholder may be annoyed at you for occupying their one free moment of the day.

Melissa was once having a difficult time getting a response from a key stakeholder. She showed up at his office just as he was walking out. He answered her first question but then said he couldn't talk and rushed off. Then she realized she was unknowingly preventing him from using the bathroom. Oops!

These approaches can border on stalking if you're not careful. It's important to try to find time with your stakeholders, but don't overutilize this approach. If you start showing up at the end of every meeting, they will certainly be annoyed, probably be a bit creeped out, and definitely start to question your motives.

An alternative to an Unmeeting is to aim lower on the organizational hierarchy. Ask the busy stakeholder if there's someone on their team that you can meet with instead. This will usually either get you a quick name (and a relieved stakeholder), or the stakeholder will realize the urgency of the request and make time to meet with you.

Irie Finally Connects with Sparks

Irie decides to take Ella's advice and find a more creative way to get a meeting with Sparks. Since she has had no success getting time on his calendar or getting him to respond to email or Slack, she decides to try to catch him in person in his office.

Irie walks by Sparks's office, but he's in a meeting, so she decides to talk to his admin, Pria, who sits just outside his office. She glances into the cubicle and finds a young, smartly dressed woman, busily typing inside. An equally young man in khakis and a rumpled blue Oxford shirt is speaking to her, saying, "I don't see how this partnership can be profitable. They need 50% of the revenue and our costs are 65%. So tell them unless we reset expectations we'd have to say 'no.'"

"Hi," Irie says in a friendly tone. Both occupants of the cubicle turn to look at her. "Sorry to interrupt. I'm Irie," she adds, offering her hand to each of them in turn. "I'm the director of product management and I need some time with Sparks." She smiles at them both but is met with blank stares.

"I am sorry to interrupt your conversation," she repeats with a nervous laugh. "You can let me know later when would be convenient for him."

"It's not that," says Pria with practiced calm. "Sparks's calendar is completely booked for the foreseeable future." Irie suddenly realizes that Sparks may be the Power Player here, but Pria is his gatekeeper, so she's a critical person in the equation.

"What's keeping him so busy?" Irie asks, curious.

"I'm not allowed to discuss it," Pria says primly. With that, Irie notices that the man in khakis seems suddenly curious. "Spill," he says to Pria.

"I'm sorry, Justin," Pria says. "It's need-to-know." Justin dramatically pantomimes being wounded. Irie decides to follow up.

"Okay, his schedule is full with something super secret. But you know his habits, right?" Irie pleads. "He must eat or go to the gym or send out for coffee," she says, reaching for any angle.

Pria reacts with a combination of pity for Irie and annoyance at Sparks. "Actually, he is really picky about coffee. He hates Starbucks as much as he hates instant coffee. He's always asking me to get coffee delivered from different local places but then he complains that it's cold by the time it gets here." Pria's complaint sounds practiced, as though she's been dealing with it so long she can't even really work up much enthusiasm for being aggravated. "He's too busy during the day to make his own coffee, but you can sometimes find him here early in the morning cursing the coffee machine in the kitchen and drinking it anyway."

This gives Irie an idea. Her native Jamaica grows some of the most aromatic coffee in the world, but only a tiny supply makes it out of the country. It is rarely fresh enough to preserve the unique flavors. Her cousin always makes sure she has a fresh stash of her favorite brew from home.

The next morning Irie shows up early with a French press and some of her cousin's beans, freshly ground at home. She heads to the open kitchen near Sparks's office and, to her luck, finds him there, cursing the office coffee machine just as Pria said. She starts to make coffee for herself with her French press, and eventually Sparks notices her.

Sparks looks enviously at her clearly superior coffee choice. Irie ignores him for a second, but then looks up and asks, "Can I make you a cup?"

As they sit together drinking the coffee, Sparks excitedly recounts a visit to Jamaica years ago and rhapsodizes about the coffee he drank there. "This cup brings me right back to Negril," he says. After a few minutes of casual conversation, Irie finds a way to weave in some questions from her interview guide. Sparks is surprisingly engaged in the conversation, despite his unresponsiveness to her efforts to meet with him. *Perhaps it is his delight at not having to deal with the office coffee machine today,* Irie thinks. She decides to abandon her stakeholder interview questions for now, and just let the relationship start to form.

After successfully connecting with Sparks, Irie is enthusiastic about Ella's Unmeeting concept. She adds a column to her Stakeholder Canvas for opportunities to connect with each stakeholder and begins recording some ideas. ■

Figure 2-8. Irie's updated Stakeholder Canvas with Connections

Stakeholder Canvas

Topic Product roadmap decisions — Product, Decision, or Initiative

Driver — Aligns with the Approver on objectives and decision criteria, seeks input from Contributors, and makes the decisions

Irie, director of product management

Objectives

Approver — Sets decision criteria, approves or rejects decisions by the Driver

Name	Role	P/A*	TBD	Connections	TBD	What They Need	What I Need
Sri	CTO	Ally					

Contributors — Add valuable information, including experts, implementers, and those significantly impacted

Name	Role	P/A*	DiSC	Connections	Opportunities	Their Needs	My Needs
Yacob	Dir. engineering	Ally	Steadiness				
José	Dir. design	Ally	Conscientiousness				
Sparks	Partnerships	Challenger	Dominant	Island coffee	Bring coffee		
Ella	CRO	Challenger	?				
Sergey	VP marketing	Observer	Conscientiousness	AI vs. people			
Divya	Dir. data science	Ally	Dominant				
Philippe	CFO	?	?				
Liandri	Dir. cust. support	Supporter	?				

Takeaways

Meeting with your stakeholders individually is very helpful in understanding their personalities, what drives their business priorities, and how they prefer to make decisions. You will get the most insight from these one-on-one sessions if you prepare in advance and take time afterward to reflect on what you've learned.

- Start with a list of key stakeholders and Power Players (see Chapter 1), and set up stakeholder interviews with them.
- Treat stakeholder interviews like customer interviews: your goal is to learn as much as you can about them, rather than to push your own agenda.
- Use what you know about how each of your stakeholders makes decisions to help you understand their perspectives and improve your ability to influence them
- Try to get meetings with high-level stakeholders, but if it's difficult to get on their schedules you may want to try an Unmeeting. It's also okay to talk with people lower down on their teams.

You may not be new to your organization but, like Irie, you can track critical information about your stakeholders in a Stakeholder Canvas for quick reference before a conversation. While these professional considerations may appear to complete your stakeholders' profiles, there is often more to the story, as Irie learned with Sergey's aversion to AI. Chapter 3, "Rapport," explains how creating a more personal connection with stakeholders can lead to deeper insight and better alignment.

You don't need to be good at politics to just be a genuine person

Chapter 3

Rapport

Now that you understand your key stakeholders better, it's time to create strong working relationships with them. You don't have to become best friends with everyone you work with, but it's surprising how much easier it is to handle a disagreement over a casual cup of coffee.

As Irie realizes that she needs to put more effort into her work relationships, we'll explore the topic of building rapport, and you'll learn about what's most important:

- Be relatable, connecting with other people at a human level and exploiting a natural human bias we have toward people similar to ourselves.
- Build mutual respect by assuming positive intent, accepting people's differences, and acting with curiosity instead of becoming defensive.
- Practice empathy by validating and sharing in someone's emotional experience.
- Encourage vulnerability—not by oversharing—but by talking about what's really going on, being willing to take on risk, and accepting the possibility of failure.

The approaches in this chapter may at first seem highly transactional, which they can be if you treat them as such. The real magic comes when you get to know people, not because you *have* to but because you *want* to. Building rapport needs to come from a place of honesty, or your actions may appear shallow or manipulative.

You will also need patience when building rapport. With some people, you may make connections very quickly, but with others it may take longer—weeks, months, sometimes years. As you work on these relationships, don't give up hope. Every step toward good rapport will make it easier to do your job.

Let's begin as Irie starts to realize that she has more work to do to gain the trust of her stakeholders.

3.1 Irie Needs to Be More Relatable

After her conversation with Sparks over Jamaican coffee, Irie realizes that a casual conversation around a shared interest—like coffee—can be surprisingly effective at getting people to let their guard down. As she discussed with Alex the previous week, sometimes jumping straight into the business of the conversation causes you to miss opportunities to get the full story from people. Irie shares the coffee story with Sri as well as some details from other, more formal meetings.

"You do make a good point," says Sri. "I've noticed that sometimes people stick to safe topics with you and don't reveal many details. They seem to be holding back. Maybe it's just because you're new, and they don't know you yet?"

"Maybe," Irie replies."You say you're 'mister no politics,' but I've noticed people tend to relax around you. What are you doing differently than I am?"

Sri sits back and takes a moment to think. "I don't think you need to be good at politics to just be a genuine person," he reflects. "People really respond when you reveal a little about yourself. I'm always talking about my latest fitness program—or my latest injury," he adds ruefully. "When it's not all just work, people tend to trust you more. And when they trust you, they're more likely to tell you how they really feel about something."

"I guess that makes sense," says Irie. "I also wouldn't trust someone until I got to know them a bit. Like when I first met Alex, I was sort of guarded because they tend to be really direct when they speak. But then we discovered that we both love a good spreadsheet and there was an immediate connection."

"Exactly," says Sri. "I don't know if this shows in our meetings, but among all the executives, I'm closest with Philippe. My guess is that's because I took up golf a couple of years ago."

"Golf?" asks Irie.

"Yeah," Sri explains. "It turns out it's not my favorite sport, but at the time I was sold on the idea that it was a sustainable life-long thing you could keep doing into your nineties. That appealed to me while I was getting over knee problems. Anyway," Sri continues, "Philippe is a huge golfer. He said he moved to the States because of all the golf courses here. He invited me to his club and coached me on my game. We got to know each other pretty well."

"And you think that's increased trust between you?" Irie asks.

"I know it has. Last time we went for funding, we talked over valuations, dilution, budgeting—all of that while walking from hole to hole on the course. I think I had a better understanding of the finances than anyone else besides him. And that helped me make smarter staffing plans, not hiring too fast or too slow. Even now," he adds, "he comes to me first to sound out next year's budget before we discuss it at Liz's staff meeting. And I run hiring plans by him first. We trust each other," he concludes, "because we have this thing in common."

Irie reflects on this story for a minute, then she says, "Maybe this is related to affinity bias."

"How so?" asks Sri.

"Affinity bias is the tendency for people to want to associate with those who are similar to themselves. Like my 'spreadsheet' example, when two people discover they have something in common, they feel a connection. This connection builds trust at a subconscious level. Though I suppose that can be dangerous," she adds, "if we only end up listening to people just like ourselves."

"But I think that's the point," says Sri. "If we want to develop rapport with a diverse group of people, we need to find those points of connection with them. They're usually in there somewhere. And if we really can't find anything in common with our colleagues, we can always create a connection, like doing some kind of team-building activity, and then we'll all have that experience in common."

Irie decides she's going to look for more personal connections with her coworkers, especially with Sparks. She and Sri discuss more ways to accomplish this. ■

Relatability

For relatability—the ability to be understood by others—the key is finding connections on a human level. This requires surfacing similarities between ourselves and our stakeholders to build effective relationships.

Affinity bias is our tendency to want to associate with those who are similar to ourselves. This can obviously have the potential for negative outcomes, such as only choosing people for your team who are just like you, leading to lack of diversity, groupthink, and stalled innovation. But affinity bias can also be leveraged to establish connections with a diverse set of colleagues. You just need to look for those points of similarity, many of which are not obvious at first glance.

Phil Hornby, a product management coach, told us that when he was a product management leader he would use his own established relationships to foster connections among other people. "When I introduced new team members around the organization, I would tell them both a bit about each other that I felt would help build a bridge, like a common interest."

Most of us have things in common, even if they take a while to discover. The best way to get a coworker to open up is to first open up yourself, to demonstrate that you're interested in building the relationship and that you're willing to go first. Keith Hopper, founder and CEO of Danger Fort Labs, talks about revealing personal details like this: "It expresses our willingness to be vulnerable, not because we're trying to get the other person to reveal something personal, but because it allows us to show up more authentically; it models the right kind of behavior that should exist between trusted partners, and it gives the other person the permission—if they choose—to also show up authentically. By moving first, we are taking a risk, but we're doing it in service of helping someone else release the weight and burden of their social armor."

Sharing personal details doesn't have to be a "tell-all exposé." It's just a little detail here and there to show that you are a real person. Instead of "I'm not available this afternoon," you might volunteer "I have a sick kid at home today." Or you might share something innocuous like "I'm going on a surfing trip, and I'm really excited about it," or "my in-laws are coming to stay this weekend, so I have to clean out the guest room."

Of course there is a "chicken and egg" problem with building trust and revealing personal details, so it's okay to start small and work up to bigger personal insights over time. Since personal details could be used to embarrass us, sharing these details demonstrates your trust in the other person: you're showing that you believe they won't use these details inappropriately. It also shows the other person that they can trust *you* to treat *their* personal details properly.

These details may also reveal that you have common interests, like skiing, or painting, or simply a love of coffee. Common interests build connections because you can relate to each other. You may even end up building a personal friendship around your shared interests. If you're both into coffee, for example, maybe you can bring in different kinds of coffee from home to taste-test at the office. If you have shared perspectives about something as simple as coffee, it can open the door to having shared perspectives about topics at work too.

When finding a connection is difficult

If your stakeholder is still not willing to share personal details, even after you have shared yours, ask simple questions. Even something as commonplace as "How was your weekend?"* can break the ice and start to build that sincere relationship that's not just about work. Being honestly curious and asking about the other person's experiences starts moving the conversation toward finding common interests. Furthermore, it shows your stakeholder that you are interested in them (and perhaps even care about them) as an individual, not just as someone you're forced to work with.

Sometimes it can feel impossible to find common ground with someone very different from yourself. Maybe you are a hyper-literate person who thrives on wordplay and clever quips, and you need to make a connection with a monosyllabic person who grew up speaking a different language and wants to stick to concrete facts. Where do you even begin? The answer is to start slowly with small things you can find you have in common, and move to bigger connections over time.

---

* Open-ended questions are much better than yes/no questions to get people to open up.

Bruce was once in a car on the way to lunch with a team from a Chinese company. They were speaking English to make him comfortable, but it was an effort. One team member volunteered that he was considering whether to move to a neighborhood they happened to be passing by at that moment. He said the schools had a good reputation. He was worried, though, that it might be too expensive.

Despite their differences in language and culture, Bruce surprised everyone by saying he thought Americans and Chinese folks were basically the same. "Why do you say that?" they asked. Bruce explained that he'd gone through the exact same train of thought when picking a location to raise his own kids. "I can't think of anything more American," he said. They all had a laugh and the lunch conversation was much more relaxed after that.

This process unfolds the most smoothly when you start identifying connections early in a relationship, but even if you've spent the last 10 years working with someone, there may still be a lot to discover. Get curious, ask some questions, and entertain the idea that you might have more in common than you previously thought. Small talk like this is expected in many cultures. We've found people vary quite a lot in how much they engage in small talk spontaneously, but you can deliberately start with small statements and get deeper over time to uncover connections (Figure 3-1).

Figure 3-1. Find an existing connection by starting small

Step	Description	Example
1. Anchor statement	An uncontroversial observation another person is likely to agree with	"It sure is hot!"
2. Personal detail	A detail specific to you that reveals how you feel about the anchor statement	"I've been waiting for it to warm up so I can go fishing."
3. Invitation	An opening for the other person to provide a similar level of personal detail about the anchor statement	"Do you have any vacation plans?"
4. Validation	A positive acknowledgment of the other person's response, possibly adding another personal detail	"The mountains? I hear the view is amazing!"
5. Connection	An observation about the connection between your personal details	"Seems like connecting with nature is something we both love."

Creating shared experiences

Sometimes you really do share very little in common with someone you work with. Even if your backgrounds and interests don't align, you can still find commonalities by *creating* shared experiences to build new connections. Depending on your comfort level and your company culture, there are different degrees of shared experiences you can create.

Figure 3-2 shows progressive steps of getting to know each other and creating shared experiences. Note that the goal is not to get to the end and become best friends with your stakeholders. The goal is to establish the right level of connection with each colleague.

Figure 3-2. Degrees of creating shared experiences

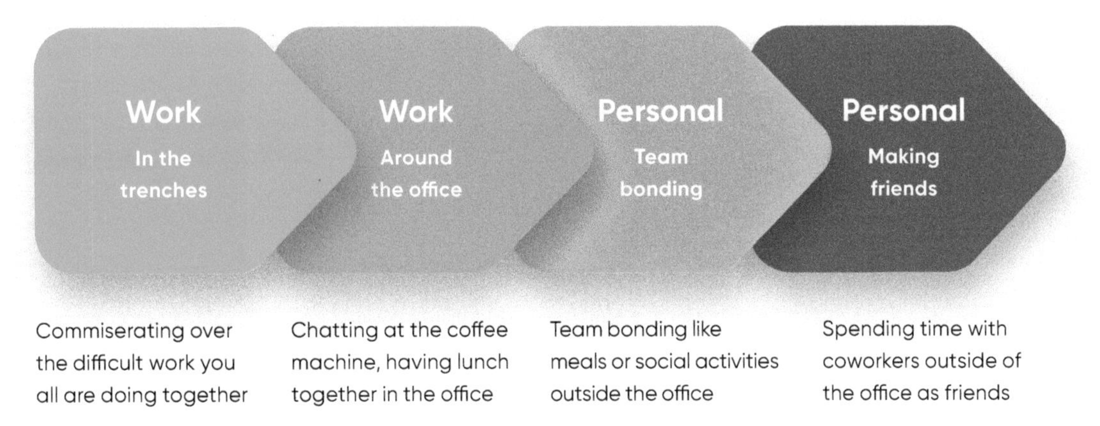

If you work closely with a small cadre of coworkers, shared experiences will come naturally from the work. Over time, you'll go through release cycles, difficult customers, painful failures, big wins, and the everyday grind of getting stuff done. You'll do this together and you'll build up a mutual understanding of what to expect from each other. But for people you don't interact with on a daily basis, there may not be as many opportunities to have shared experiences happen naturally. You'll have to manufacture them.

Creating shared experiences goes a long way toward creating well-functioning, productive stakeholder relationships. There is a reason why companies sponsor offsite events, retreats, and conferences. They all have official agendas and plans for specific outputs, but their biggest utility is in creating a shared experience outside the office routine. We all put on a mask of professionalism at work and, importantly, we all *know that* about each other.

Outside the carefully managed office environment (in person or virtual), we tend to act more like ourselves: more authentic, more like real people, not just coworkers. Allowing others the opportunity to peer behind your mask to see the real you makes you more relatable. Some people may worry that having too many "coffee chats" will be seen as a waste of time, especially in a virtual environment. But creating casual time to connect, even virtually, goes a long way to building rapport with your coworkers.

You don't have to wait for a big conference to create these shared experiences. You can do something simple like inviting your engineering team to lunch and inviting a stakeholder or two to join you. Or invite your stakeholder to travel with you to a customer site, which has the dual benefits of the stakeholder gaining valuable customer insight, and having a few social dinners while traveling. Or, especially with post-pandemic hybrid-remote working models, just meeting someone you don't see every day in person for a walking one-on-one is extremely helpful to build rapport and to get a glimpse into each other's authentic selves.

At one company, Melissa decided that there were people in her broader team with whom she rarely connected. So she created a monthly "Women in Warehousing" brown bag lunch in the office lunch room, for all the ladies in the department to get together. There was no agenda, just an opportunity to connect casually and build rapport. As conversations naturally incorporated work-related banter too, Melissa often learned valuable new information, leading to follow-up conversations and connection building with some of the women in the group.

"The act of connecting with others puts everyone at the same level and breaks down any hierarchies and titles. We're all people; people like to be heard and feel special. While fresh new interns cannot add much when discussing work-related topics with CEOs or senior leaders, when talking about themselves, they are the experts on the subject. So, talking about themselves boosts their confidence."

Pedro Amaral, vice president of product

Some ideas to create casual environments to connect:

In person

- Go for a walk and/or get coffee.
- Have a "walking one-on-one" meeting.
- Have lunch in the lunch room instead of at your desk.
- Do some work in a common space in the office and see who comes by to say hi.

Virtual

- Schedule virtual "coffee chats" (one-off or recurring).
- Online team-building activities, like Jackbox games.
- Weekly virtual team happy hour after work—no agenda, just chitchat.

Joseph Spooner, an IT specialist at the California Department of Technology, recalls one brilliant colleague who no one could get along with. People were constantly fighting with him, so Joe decided to take action.

He decided to go for a "walk and talk" with this colleague, not to give him a lecture, but to get to know him. "By doing that, I'm actually engaging some of the sympathetic nervous system. Walking causes them to breathe normally, and regulate their thoughts a bit more. It gives them a chance to pause and think, and they're actually stimulated by the outside environment while having this conversation.

"As I talked to him, I realized he just felt so insecure. Whenever he felt challenged, he felt the need to challenge back because he had a military background."

Joseph points out that it's a useful technique to literally get out of the boxes we're all sitting in and get to know each other personally. An additional advantage of walking is that you're not physically facing off against each other because you are facing in the same direction. "Almost every time I've done this, the walls come down," Joseph continues. "We have a more meaningful conversation, and that starts to create trust, reducing insecurity. And if that person is the one influencing the rest of the group, I've made a lot of progress in an hour or two."

Irie Makes Connections

Over the next few weeks, Irie tries to build rapport with her stakeholders by beginning all her meetings with a casual question or observation. She finds this is a surprisingly easy way to uncover useful details about people and make connections.

She's been at Helthex almost two months now, so she's met most of her stakeholders already, but she's still learning new things about people. A comment about the weather reveals that she and Divya, director of data science, both make a ritual of kayaking in nearby lakes every spring. She learns that Zola, the customer support manager, is getting certified as a yoga instructor and hopes to open a studio someday. These more personal conversations make the business discussions easier.

Irie updates the "connections" column in her Stakeholder Canvas as she discovers connections.

Irie has been avoiding Sergey, worried she might offend him again after he stormed out of their first meeting. She needs to speak with him about the messaging for a new feature they plan to launch next month, however, so she decides now is the right time to try again to build a good relationship.

Irie starts the meeting with Sergey—this time a video call—more casually than last time. "Any plans for the weekend?" she asks. Sergey replies that if the weather holds, he is planning to have his brother and his family over for a cookout. Seizing the opportunity, she says, "I heard about your brother's job. I'm sorry."

"No, I'm sorry," says Sergey. "I overreacted the last time we met. AI is here to stay and we need to learn how best to work with it. I'm helping my brother try to get a new job, and we've been learning about AI together over the past few weeks."

"So you wouldn't object if we looked into how it might help with our recommendations?" Irie asks, just to be sure.

Sergey crosses his arms protectively. "As long as we are careful about accuracy, safety, and liability," he says. "And as long as we are using it to add new value, not replace people." He pauses. "I don't mean to bring my personal biases into the conversation."

"Actually, I'm glad you have a conscience," Irie says, smiling. "I think a lot of people agree with you about the proper uses of AI, and that should be part of our decision."

"Exactly," says Sergey. "I'm no technical expert, but I'd love to review any product ideas you have. I have to sell it to the market, and I want to believe in what I am selling."

"I'm happy to include you earlier in the process," says Irie, "so you can give us some input from a market perspective. And from a personal perspective."

"Happy to help," says Sergey. ■

3.2 Irie Adds Respect to Her List

That afternoon, Irie sees Sri in the hallway. "Walk with me?" asks Sri. As they walk toward Sri's next meeting, he asks Irie about how her adventures in relatability are going. He also mentions that Sparks set up a meeting with him about the AI topic.

"I'm glad he's thinking about the topic more," says Irie, "but why is he setting up a meeting with you instead of me? I thought I was doing better with him."

"Don't worry about it," says Sri. "We'll get there. I'll invite you to the meeting too, and I'll let you do most of the talking. I think it's just because Sparks still doesn't know you that well."

"I guess I still haven't earned his respect," says Irie.

Sri pauses outside the door to his meeting room and brings his voice down to a whisper. "Sparks doesn't respect a lot of people. I'm honestly not sure how I got on the list. He can be so difficult, but I show him respect, even when he doesn't deserve it."

"Maybe that has something to do with it," Irie says. "Maybe part of earning respect is showing it. You know, giving the other person the benefit of the doubt on something even if it doesn't seem to make sense at first?"

Sri thinks a moment before answering. "Years ago–when I hadn't been here very long–Sparks came to me with this idea of adding a voice memo feature to the app. I thought it was crazy but I heard him out. It turned out he thought of it as a way of collecting data that could be easier than typing or clicking buttons for some people. We didn't build exactly what he described, but it's become a popular way to record symptoms throughout the day. Divya even thinks she could use transcriptions of these recordings to feed into our models. In the end I was glad I listened," adds Sri. "Maybe he remembers that."

"That's a great story, Sri," says Irie. "You wouldn't have discovered the valuable kernel in his idea if you hadn't followed up and listened a bit longer to get at it."

Irie thinks about this advice as Sri enters the meeting room. ■

Respect

Showing respect is an important way to build connection and rapport with others. Respecting others means treating them with due regard for their feelings, wishes, and rights. To do this, you must accept their individuality and beliefs, regardless of whether you share the same viewpoints or values.*

Respect is a basic human need, and it's impossible to form a real connection without respecting each other. Different cultures may have different ways of showing respect, and you should learn those differences if you work with cultures other than your own. But many of the same expectations hold across cultures. Examples of behaviors that demonstrate respect are being kind, considering how the other person will hear your words and choosing them carefully, not interrupting, avoiding gossip, and being on time and prepared for meetings.

Here again, curiosity is important. Building rapport with someone means getting to know them, by being honestly curious about them. Even if the other person's opinion is different from yours, they likely have entirely valid reasons for feeling the way they do. By finding out what those reasons are, you may even find that you have more in common than you thought. You may sometimes even change your own opinion.

Assume positive intent

Assuming positive intent conveys respect because it assumes that the other person has valid reasons for their beliefs and actions. When you see a behavior or hear a statement that doesn't make sense to you, you should assume there is something that would make their behavior or opinion sensible—you just don't know what it is yet. The next step is to try to discover what it is that is causing the puzzling behavior, rather than judging the person. The missing link might be something they don't know. Often, it is something you don't know.

Disrespect can foster team distrust, infighting, and reduced team efficiency. For example, Melissa was coaching two teams who were not working well together. In a workshop, the engineering manager from

---

* We discuss ways of building respect for your own expertise and credibility in Chapter 4, "Trust."

Team 1 made a proposal about a new team organization model that the engineering manager from Team 2 did not like. The proposal suggested splitting into three engineering teams: one for feature work, one for customer requests, and one for fixing bugs.

Team 2's engineering manager told his colleague from Team 1 that his proposal "goes against everything I believe in as an engineer," and then went into a five-minute rant about how terrible the idea was. This outburst showed a lack of respect because it did not demonstrate any curiosity about the intentions behind Team 1's engineering manager's proposal.

Melissa didn't like the idea either, but, assuming positive intent, she said, "That's an interesting proposal. It sounds like the problem you're trying to solve is that the engineers are getting distracted by frequent customer requests and bug reports, and are having a hard time delivering on the roadmap." Team 1's engineering manager said, "Yes! That's the problem I'm trying to solve."

Eventually, Team 2's engineering manager agreed that distraction of the engineers was a valid problem to solve, and the two engineering managers were able to work together to come up with an alternate solution following the intent, but not the exact design, of the original proposal.

Value people's time

Another important aspect of respect, especially when dealing with stakeholders who have a higher rank than you, is valuing their time. Of course this respect should be mutual, and everyone should value other people's time, but start with what you can control—your own behavior.

Showing respect for people's time can be exhibited by showing up on time to meetings, having a clear agenda, and sending reading materials in advance of a meeting. It's also about responding promptly to emails or other messages. You never want a stakeholder to have to repeatedly ask for answers to their questions or to go to your manager when they don't hear back from you. This doesn't mean that you have to provide the answer right away, or be glued to Slack and email 24/7, but if you see a question from a critical stakeholder, it's easy to type out a quick response that you got their message and you are working on it.

You can also value their time by providing short, concise answers to their questions. Long, winding answers and explanations can leave senior stakeholders wondering why you're wasting their time. It's worth a bit of *your* time to work out in advance some simple ways of explaining things to stakeholders, and to clearly articulate what you need from them.

Respect goes both ways

What do you do if you're being respectful of your stakeholder but they're not being respectful of you or your time? Clearly, there are some forms of disrespect that should not be tolerated, like insults, use of offensive language, or sexual harassment (all of which would warrant a report to HR), but there are other, milder forms of disrespect that can be tackled by trying to better understand where these behaviors are coming from.

For example, if your stakeholder is micromanaging you, it's possible they need more frequent communication on the status of the project.* Excluding you from certain meetings might mean your stakeholder doesn't understand your role on the team. A stakeholder who constantly interrupts others in meetings might benefit from a more concrete meeting agenda and some feedback about the benefit of ensuring that more voices in the room are heard.

* For a useful perspective on the relationship between trust and communication, see Emily Levada's article, "The Trust-Communication Trade-Off," October 11, 2018. *https://medium.com/@elevada/the-trust-communication-trade-off-4238993e2da4*.

Irie Practices Respect

The meeting on AI with Sparks is scheduled for the following Friday morning. Irie and Sri are each remote as their teams work from home on Mondays and Fridays. They chat while they wait for Sparks. Eventually Sparks joins the meeting from his desk at the office.

"I wasn't expecting you at this meeting, Irie," Sparks says as his only greeting.

We're off to a great start, thinks Irie. Calmly, she responds. "Sri thought it made sense for me to join, since I'll be owning the product if we end up adding AI to it."

"You mean *when* we add AI," challenges Sparks.

"Okay, you two, let's have a civil discussion here," interrupts Sri.

Irie realizes that she is already off to a bad start. And she's missed her opportunity to make some small talk with Sparks—to try to find some way to relate to him—but it is too late for that now. "I did mean 'when.' Sorry I misspoke," Irie says. "I'd like to talk about what it would look like to add the AI features you're suggesting, and the trade-offs we'd have to consider to make it happen."

"That's better," Sparks begins. "Because the board is really keen on getting AI into the product. It was my idea and they are super excited about it. It's just the thing to get us more funding. They have even introduced me to a few other portfolio companies to talk about..." here he pauses as if uncertain what to say. "...partnerships."

"We should definitely look into it more. Can you tell me a bit more about your specific ideas?"

"I don't have the ideas, Irie, that's your job," Sparks snaps.

Sri jumps in. "Sparks, you don't have to yell, she was only asking if you have any ideas."

Irie follows up quickly before Sparks has a chance to get angry. "I will work with my team to do some research and see what we can come up with." She pauses and then takes the opportunity to ask for a one-on-one meeting with Sparks. "I'll work with Pria to set up some time on your calendar to review. I'll CC you on the email so she knows you okayed it. Does that work for you?"

"Fine," says Sparks. And he leaves the meeting.

Still on the call, Irie and Sri are a bit stunned. "Do meetings with Sparks always go that well?" asks Irie.

"Usually," says Sri. "Like I said before, he doesn't respect other people. But you did great. You kept it cordial, took ownership, and offered a follow-up plan. Don't let him get to you. Sometimes we just have to deal with difficult people."

"I thought our bonding over coffee the other day would make him nicer," says Irie, "but he's really unpredictable." Irie reflects for a second on what Sparks said. "And now I'm really curious about why he's so passionate about this AI thing. He gave us some clues when he mentioned the board, and that it was 'his idea.' I'm going to follow those leads and see what I can come up with. It feels like I'm missing something." ■

3.3 Irie Tries to Take Sparks's Perspective

After the disastrous call with Sparks, Irie tries to find out more about the board's perspective on AI, and why Sparks is so fascinated by it. She starts with Philippe, the CFO, because he might know more about the funding that Sparks mentioned. Irie schedules some time with him, calling the meeting "Intros Irie/Philippe" and writing in the description "Hi Philippe, I'm Irie, I just started as director of product management, working for Sri. I'd like to introduce myself and discuss how our teams can work together." Luckily, Philippe accepts the invite.

Philippe stands and reaches over the table to shake Irie's hand as she enters the small conference room. "Great to meet you, Irie. I've heard a lot about you from Sri."

Irie explains that she's trying to get more context around the AI work that Sparks is asking for. She also finds a way to mention that she's gained a bit of a following when she brings her Jamaican coffee to the office.

"So that's what the party's all about in the mornings," says Philippe. "I am more of a tea drinker. Which is strange, since I am from Paris originally. But I guess I just don't like going with what everyone else is doing."

"Me neither," says Irie. "You gotta keep them guessing." They share a laugh and then Irie continues. "I'd really like to hear about your role as CFO. I'm curious how you think about product management relative to the work you're doing."

"Interesting question," says Philippe. "My understanding is that your job is to help us create and launch new features in the product. Is that right?"

"That's a part of it," Irie replies. "Our job in product management is to solve for what customers find valuable, what is technically feasible, and what meets our business interests as well."

"I didn't realize you looked at the business side, too," Philippe says.

"Well, it's no use spending months building something—even if customers love it—only to discover we can't make money on it."

"I completely agree," says Philippe. "And now I am very glad we hired you! I sometimes feel like my arguments about profit and loss evaporate after they leave my mouth. No one seems to be listening."

"I'm glad we are talking about this, too," says Irie. She goes on to ask him about his view of the proposed push into AI.

"Well," explains Philippe, "Sparks has been pitching an AI concept to the board for a while now. The board is a bit hesitant, but Sparks thinks it's the ticket we need to differentiate ourselves from the market and attract partners to work with us, which could help us make the case for more funding. That funding is clearly interesting to me as CFO."

Irie pauses to consider this. It sounds like the board is not totally on board with the AI concept, but Sparks made it sound like they were very supportive. "And what do you think about AI?" she asks Philippe.

"I'm not sure that AI is the 'golden ticket' Sparks says it is, but it sure seems like something a lot of people are pursuing these days. I'm not in technology, so maybe you can tell me if it's worth investing in?"

"I'm trying to establish exactly that," Irie explains. "It's really helpful to have your perspective."

Irie has a one-on-one that afternoon with Yacob, her peer in engineering. She tells Yacob that she promised Sparks she'd look into how AI can be worked into the product, and that she needs his help to do the research.

"It's great that you're doing that, Irie, but I just don't have time. We're trying to finish the architectural design for the new platform, and my top engineer quit on me yesterday. It's been hard to hold the team together when the strategic direction keeps changing. And now we're adding AI to the mix? Feels like we can't make up our minds."

"I hear you, Yacob. I don't like changing direction all the time either. But Sparks seems really passionate about this AI idea, and he isn't going to drop it. I think we should at least look into it. If it makes sense to do, we'll do it. If not, we'll have good reasons to reject it."

"I guess. I believe you have our best interests in mind, but this is Sparks's usual game. He has us go after whatever is his 'idea of the week,' as we call it. It's really demotivating for the team. That's part of the reason that my engineer quit."

"Wow, I didn't realize it was that bad," says Irie, thoughtfully. "For now, we don't need to involve the team. I don't even know if this idea is going to go anywhere. And I know you're short on time, so I'll start looking into it on my own. But can I bring you in if there's a technical question?"

"Of course. And you should also connect with Divya, because she's head of data science."

"Right," says Irie. "We haven't had a chance to talk about AI."

"And I didn't mean to say I wouldn't help," Yacob clarifies. "I guess I'm just frustrated, that's all. I appreciate you letting me vent."

Irie smiles and tells Yacob not to worry about it. "We are all just people doing the best we can."

Irie decides she's going to figure out why Sparks is so interested in AI by trying to see things from his point of view. First she will find out as much as she can from the people in his circle, using stakeholder interviews specifically about the AI topic. Maybe this will help her to understand Sparks's perspective. ■

Empathy

Empathy is difficult. It involves stepping outside our own heads to see things from other people's perspectives. It's natural, and evolutionarily appropriate, to be self-centered and think first about our own needs and perspectives, which is why it's important to remind ourselves constantly to consider what other people might be thinking and feeling.

In his classic book, *How to Win Friends and Influence People* (Simon & Schuster, 1936), Dale Carnegie says that "success in dealing with people depends on a sympathetic grasp of the other person's viewpoint." Although Carnegie used the word "sympathy" in his writing almost 100 years ago, his usage matches closer with the modern meaning of the word "empathy."

Empathy is the ability to understand and share the feelings of another person. The *Merriam-Webster* online dictionary makes this distinction: "While sympathy is a feeling of sincere concern for someone who is experiencing something difficult or painful, empathy involves actively sharing in the emotional experience of the other person." Empathy is being in the moment and feeling someone's emotions with them. Sympathy, on the other hand, is more of a sense of pity for the person, without a true connection to or acknowledgment of the emotions they are feeling, like "oh, you poor thing."

Developing empathy with your stakeholders can be a powerful way to build rapport. Understanding your stakeholder's perspective usually involves a bit of research, but once you have an understanding of your stakeholder's point of view, incentives, and success metrics, you can use that information to build a better connection.

Chapter 2 provides some tips for stakeholder interviews. Expanding on that, next we describe four valuable conversational techniques that you can use in any situation to demonstrate empathy and connect with your stakeholder (or anyone else): open-ended questions, active listening, mirroring, and summarizing.

Open-Ended Questions

An open-ended question is one that invites a longer, descriptive answer, not a yes/no or numerical answer. This type of question also invites opinion and explanation, free from constraint. Instead of "Does this affect your team?" (a yes/no question) try "How does this affect your team?" You can also follow up on statements they've made for more information, like "Please tell me more about that."

Active Listening

Active listening means that you are absorbing what the person is saying, not just waiting for your turn to speak. It's also about demonstrating to the other person that you are really listening. It can be difficult to give your full attention to someone else, rather than thinking about how the conversation relates to your own experiences, but it becomes easier with practice. Here are some ways to demonstrate active listening:

- Nod, make small verbal acknowledgments like "Mmhmm," "Right," or "Ah," or laugh or smile at the funny bits.
- Acknowledge the importance of what they are sharing by showing empathy, using phrases like "What an amazing accomplishment!" or "You must have been scared."
- Ask follow-up questions, such as "How did that happen?" or "What's the effect on your team?" or "What happened after?"
- Give them a few extra seconds after they finish speaking, which often prompts them to volunteer more on their own.

Mirroring

Mirroring is repeating or duplicating someone's words, speaking style, jargon, tone, or posture, in order to validate their position, or prompt them to share more information.

For example, if they use a regional or nonstandard term (like "pop" vs. "soda"), use their term to demonstrate familiarity. Note: be careful not to inadvertently mimic someone's accent—they could perceive it as mocking.

If they are not forthcoming with details, you can mirror by repeating key words they've said in the form of a question to get them to say more. Do this in an empathetic or curious manner. For example, if they say "Customers don't like this feature," you can say "Customers don't like this feature?" and they'll naturally expand on their thoughts.

Physical mirroring can include taking the same posture as them, or using physical gestures to express the same emotion, like excitement or boredom. Sit if they sit, stand if they stand, lean in if they lean in. If they seem excited and start talking faster, you can do the same to show that you are also excited about the idea.

Summarizing

Summarizing is a way to validate your understanding by restating what you think the other person said in your own words. When you get it right, this also has the effect of convincing the other person that you empathize. When you get it wrong, you give them an opportunity to correct your misunderstanding.

The goal is to get to a point where they say something like "Yes, exactly," or "That's right." Then you and they both know you understand what they mean. For example, you could say, "It sounds like what you're saying is that you'd like to help, but your blocker is that you don't have enough people on your team to do the work, and you have higher-priority initiatives, is that right?"

Phrases like "it sounds like..." and "it seems like..." are helpful in presenting a neutral reiteration of what someone has said, to look for confirmation (or correction) without judgment.

In *Never Split the Difference* (HarperCollins 2016) Chris Voss calls these types of statements "labels" and finds them helpful to get someone talking and revealing more information. He further recommends being quiet and listening to the answer.

Irie Demonstrates Empathy

Irie spends the next week talking with stakeholders around the company. She talks with Divya, director of data science, and discovers that she has a number of ideas in the AI space, but she's not sure there's a market for them. Sergey, VP of marketing, confirms he is willing to collaborate on market research and messaging.

Irie chats with Ella, the CRO, and finds out that customers are interested in AI because they've heard about it, but it's not a deal breaker, particularly with corporate opportunities. "They are being cautious about adopting AI internally because of the risks," Ella explains. "They are much more insistent about admin features, compliance, and privacy."

Irie detects a note of resignation in Ella's voice and she remembers what she said in their first meeting about never seeming to catch up on corporate feature requests. "This is weighing on you," Irie observes.

"It really is," Ella replies. "It's hard fighting a lot of battles you know you will lose."

"I used to be an engineer," Irie shares. "In my first real job I was assigned to a product that we were planning to replace. They were building a new product but I had the task of squashing bugs in the old one. It was so flaky after so many years, though, it was an impossible task. Every time I fixed one thing, three other things would break. It was like whack-a-mole. And the worst part was that I knew we would be shutting the thing down."

"That feels exactly like where we are with selling to big companies," Ella says. "To be honest," she adds, "I'm not sure we should be going after corporations, at least not with the product as it is now. I think it distracts from our consumer efforts."

Irie is surprised that Ella is pursuing a sales strategy she doesn't believe in. As CRO, it seems like she should have the authority to change it. But then she remembers the company's Directive decision-making culture. Irie asks how they came to pursue this strategy, and Ella admits that Sparks made the decision. "He convinced Liz to fund a sales team for large customers. Arianna runs that team. She reports to me, but Sparks hired her."

After her conversations with Ella and the other stakeholders, Irie reflects that when she shares her own experiences, others are more willing to open up to her. This technique has given her a much deeper perspective on how decisions are made in the company. A lot of it seems to come back to Sparks.

Irie also speaks with Christina, the PM in charge of analytics, and she has the same perspective. To her, AI feels like Sparks's latest "idea of the week." With the help of her team and stakeholders, Irie has nonetheless been assembling ideas for adding AI into the product. As these efforts continue, Irie decides she needs to understand what's in it for Sparks.

Irie shows up at the office early with coffee, as she's done before. Sure enough, the pleasing aroma brings Sparks out of his office. He sits with her as she pours him a cup. The last time they spoke, Sparks was very curt. Happily, he seems to relax a bit with the coffee.

Irie begins, "I've been looking at some different AI ideas. I think we can find a few that will deliver some value. But it's not a slam dunk. What would it mean to you personally if this doesn't work?"

Sparks thinks for a moment. "I'm not sure anyone has asked me that." He pauses before continuing. "It's important for this company to have something that differentiates us from our competitors. Otherwise we're vulnerable. I don't think that's only for me, that's for everyone."

"Yeah, differentiating ourselves makes a ton of sense," says Irie, summarizing. She pauses to see if Sparks will keep talking. He does.

"And we really need a new idea, something that sets us apart. We need innovation."

Irie doesn't like the word "innovation," because it's often used as a generic term for "make more stuff," but she gives Sparks the benefit of the doubt and lets him continue.

"My father cofounded a networking equipment company back in the day," says Sparks. "I worked there summers as a technician. Liz was a product management intern. That's how we met."

Irie laughs and says, "You have known her a long time."

Sparks continues with his story, not seeming to hear. "The company grew like crazy for a while and we had all kinds of money. That's how I got to visit Jamaica and St. John, and St. Lucia, and all those islands while I was still in school.

"It didn't last, though," Sparks continues. "There was a downturn and there were layoffs and my dad and his partners ended up selling the company to Cisco for cheap. He still walked away with a decent amount of money, but he was always disappointed with that. He made a lot of angel investments but nothing ever panned out. He gradually lost all of it. He gets a check from Barney every month."

"Barney?" Irie asks.

"Oh, my real name is Barnabas," Sparks confesses. "It's a Greek family name. I was named after my grandfather, but my parents call me 'Barney.'"

Wanting to extend this moment of intimacy and calm, Irie asks how he got the nickname "Sparks."

"Liz gave it to me, actually," he says, smiling for the first time in Irie's experience. "When we worked for my father, I overclocked some of our devices to see what they could do and one of them threw actual sparks. Liz was sort of making fun of me by calling me that. Some people laughed but it was better than the references to purple dinosaurs I got in school. So I kind of adopted the nickname."

Irie chuckles and then tries to summarize. "So, for you, innovation is how we avoid the company 'selling for cheap,'" Irie says.

"Yeah," says Sparks, "If we do something big, we either grow or we attract a lot of investment. Either way, we're in a better spot."

"This is very helpful," says Irie. "Thank you for sharing. Let me apply this lens to the ideas we've been developing and let's talk in a few days." ■

3.4 Irie Encourages Vulnerability

The next morning, Irie gets an early chat message from Eitan, the PM on her team in charge of admin tools for the app. Eitan asks Irie if she can join a meeting already in progress with him and Wei, a lead engineer. Irie logs on, and discovers that Eitan and Wei are having an argument about which direction to go with a particular feature.

"I really don't see why talking to one customer means that we have to completely change the way this feature works," says Wei.

"It wasn't one customer, it was eight, and they all said the same thing," says Eitan. "Besides, the change isn't that big."

"It's definitely big," says Wei. "This change is going to set us back two weeks!"

"Irie, what do you think?" asks Eitan.

"Wow," says Irie, still a little sleepy. "Can we please back up so I can have some more context? I'm not familiar with this feature."

Eitan explains the feature to Irie. Then he explains the problem he saw when talking with customers. "We got several requests to consolidate all of these steps onto one screen," he explains. "I talked to the people who made the request and they all said it would speed up data entry if they didn't have to pause to move to the next screen between steps. It makes perfect sense." Wei sits back with her arms crossed and doesn't say anything. She seems distracted.

"And what's more," says Eitan, "I recorded the meetings so you can see how confused the customers are about why we make them go through all this navigation."

"What about you, Wei?" Irie asks. "What's your perspective?"

"I just don't understand why the customers want to use the product that way. We designed it to explain each step individually to ensure users understand what's needed." she replies.

"I don't understand either," says Eitan. "But this is what they're doing, so we have to adapt."

"Have you asked them why they use the workflow that way?" asks Wei.

"Well, no, I didn't have time," admits Eitan, with some embarrassment. "Maybe they don't need the explanation?"

"I've been in this situation before," adds Wei, without criticizing Eitan's mistake. "Sometimes they don't actually want to change the feature they're using—they want a completely different feature. And other customers want to leave the original feature alone. Changing a feature for people who are using it for an unintended purpose can mess things up for the customers who are happy with the original."

"That would be particularly true here," Irie speculates, "if these were experienced users who don't need the detailed explanation on each screen, while novice users do."

Eitan takes this in. "Okay, well, I could go back to the eight customers I already talked to and ask some follow-up questions. I could also recruit some newer customers to interview for a more complete picture. Maybe you can help me with the questions, Wei, since you've seen this before. José's team is always really busy, so maybe we can do the interviews faster if we split them up."

"Happy to help however I can," says Wei, smiling now.

"Sounds like you two were able to work this out without me," says Irie, pleased. "Let me know how it goes."

After she leaves the meeting Irie reflects on what she heard. *Eitan admitted that he didn't ask certain questions, instead of trying to prove he was right,* she says to herself. *He even asked Wei for help. Wei took his willingness to be vulnerable really well. She didn't yell at him about not knowing what he was doing. Instead, she welcomed the opportunity to help make it better. Maybe I can use that technique with Sparks.* ■

Vulnerability

In *Dare to Lead* (Random House, 2018), Brené Brown defines vulnerability as "the emotion that we experience during times of uncertainty, risk, and emotional exposure." She further describes vulnerability as "having the courage to show up when you can't control the outcome." Vulnerability is not disclosing everything about yourself or oversharing. It's openly acknowledging what's really going on, being willing to take on risk, and accepting the possibility of failure.

When people are comfortable being vulnerable with each other, they are more willing to say what they really think. Instead of saying what feels safe, expected, and noncontroversial, you are more likely to get "the real story." This can prevent a lot of misunderstandings and reveal hidden misalignments.

Melissa once had a stakeholder who wanted to expand a product launch to smaller facilities. The stakeholder thought the small facilities were similar enough to large ones, but Melissa was concerned the product was too complex for their operations. The two argued until Melissa admitted that her opinion was based only on personal observations. She had not asked the GM for the small facilities for his opinion.

When Melissa admitted this, the stakeholder admitted the same thing: he had not spoken with the GM either. Melissa then mediated a conversation between the stakeholder and the GM, and they collectively agreed not to launch the product in the smaller facilities.

As this example shows, one way to encourage vulnerability in other people is to start by being vulnerable yourself. For example, you can admit you are feeling overwhelmed, or be frank about your opinion of the corporate strategy. Taking a risk and saying something when you're not sure how the other person will react will get them to feel comfortable that they can do the same with you.

Figure 3-3 shows some examples of ways you can demonstrate your own vulnerability.

Figure 3-3. Examples of statements that demonstrate vulnerability

Type of statement	Example
Unpopular opinion	"I know everyone said they liked the COO's presentation this morning, but I actually think he was understating the magnitude of the problem."
Feelings and emotions	"When you told me what you thought about the status update, it felt to me like you were saying our team wasn't trying hard enough."
Personal insecurity	"To be honest, I'm not sure what the best solution is, and I'm struggling to find the right framework. Can we talk through it together?"
Admitting mistakes	"I admit that I hadn't considered that particular risk. We have a solution now, but I'm going to go back and figure out how we missed it."
Asking for help	"To be honest, I have a lot on my plate right now and I'm struggling to find time for this project. Do you think anyone on your team would be willing to help me with it?"

Psychological safety

To create a psychologically safe space, you must free people from worrying about the repercussions of stating an opinion, sharing how they feel, or making a mistake. This will give them the freedom to be vulnerable.

For example, if someone states an opinion or a feeling that offends or confuses you, you might be tempted to react negatively or defensively, perhaps telling them why they're wrong. But if you reprimand them for their opinion, it's unlikely that they will share another opinion with you in the future. They may even withhold other information from you for fear of how you will react. Instead, ask follow-up questions, like "What makes you feel that way?"

"What" and "how" questions are better than "why" questions because "why" sounds judgmental, even accusatory. We can word the same questions in different ways and get different results. Look at the questions in Figure 3-4 and think about how you would feel if someone asked you each one.

Figure 3-4. Examples of more and less judgmental versions of the same question

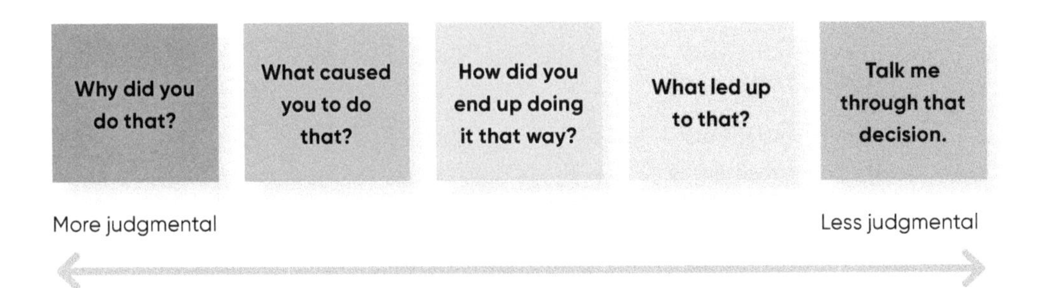

Some companies and teams do not foster psychological safety, and you may not have control over the culture at your company. But you can create safe spaces on a one-on-one basis with your stakeholders by being curious about how they feel, and then reacting in a way that encourages them to be honest.

Many Agile teams already have a regular ceremony designed to help the team improve by encouraging open discussion: the "retrospective" or "retro." The retro provides an opportunity for the team to check in with each other periodically. It's intended to create a safe space for team members to share how they're feeling about working on the team, to identify problems, and to develop possible solutions to those problems.

A retrospective ceremony is a meeting where the Product Team reflects on the most recent sprint, or on an incident that happened with the product, and asks themselves what went well and what could have gone better. It is an opportunity to expose problems on the team in a psychologically safe way that allows the team to address the issues and improve the way the team works. Often managers will not attend the meeting, to allow the team members to speak more freely.

Effective retros allow the team to talk about what happened and not who gets credit or blame. As Norman Kerth says in *Project Retrospectives* (Dorset House, 2001), "Regardless of what we discover, we understand and truly believe that everyone did the best job they could, given what they knew at the time, their skills and abilities, the resources available, and the situation at hand." If you focus on learning rather than assigning blame, you can avoid repeating mistakes in the future.

Ask for feedback

Asking for feedback from stakeholders is a great way to create a space for them to share how they're feeling. You can use a formal method, like a retro, or an informal method, like asking simply, "How do you think that meeting went?" Periodically, say once a month or once a quarter, you can ask stakeholders, "How do you think the project is going?" You can also ask some follow-up questions like, "Is there anything you think we should be doing differently?" or "Do you have any feedback for me?"

This will often prompt your stakeholder to ask the same questions back to you. If they don't ask for your feedback in return, you can be more direct. For example, "I was wondering if we could talk about the meeting yesterday. I felt like we were not on the same page, but I'd like to hear your perspective." Or "I've been getting a lot of requests from your team lately. Has something changed with your strategy or have you given the team some new objectives? What am I not seeing?"

By focusing your questions on understanding, and assuming positive intent, you can approach these important topics while being respectful and signaling that you are open to the stakeholder's perspective. Stakeholders are often relieved that you've brought up a topic they were afraid to mention. They may even thank you for highlighting where they haven't been transparent enough with you.

Irie Shows Vulnerability

Later that week, Irie shares with Sparks an update on her efforts to develop AI ideas, as promised.

"I hope the board likes these ideas," says Sparks. Then in an angrier tone, he adds, "And they better be good ideas." He pauses, sits back, and speaks softer now, even though there's no one eavesdropping. "I'm the key representative of this company to the board, so these features need to work, or we'll lose face."

Irie takes a moment to consider what Sparks is saying. When they spoke last, he opened up about his father's failed business, and she came away thinking they'd made a connection. She thought she understood that he was worried about financial security for the company and for himself. Now, though, he is talking about his increased authority while Liz is out on medical leave. "How does Liz fit in here?" asks Irie.

"Of course Liz is the CEO. But while she's out, I'm in charge." Sparks leans forward and lowers his voice a little. "It's important. The board trusts me, which is why they've asked me to present to them. Liz trusts me a lot too." He sits back. "It's a big deal to get it right."

Irie isn't sure if he's bragging or complaining, but it seems like Sparks is anxious about doing a good job. Irie decides to take advantage of the situation and open up a bit herself. Sparks's comments about the risk to his reputation if the AI idea doesn't work are the first glimpse Irie's had into any self-doubt from him, so she decides to be vulnerable too.

"Yeah, I've been there," says Irie. "A few years ago I had to make a presentation to the board of my company and I totally messed it up. My slides didn't go over well, and people were confused. Plus I wasn't able to answer several of their questions. My boss said he wasn't upset with me, but it took a while before I was allowed in front of them again."

"Well, that's not surprising," says Sparks. "I bet your boss let you talk to the board when you weren't ready yet. Too junior."

"Maybe," said Irie. "But I learned a lot, and the next time I had a chance to speak to the board I was much better prepared. The experience definitely made me a better storyteller."

Sparks sips his coffee and thinks for a minute, and then says offhandedly, "If you're such a good storyteller, maybe you can help me tell the right story about AI to the board."

"I can definitely do that," says Irie, sitting up straight, and hopeful about the opportunity.

"Okay, I'll think about it," says Sparks quickly, seeming suddenly to realize he made the proposal out loud. "I have another meeting starting now," he adds as he turns to his computer and motions Irie out of the office with a flick of his wrist. ■

Takeaways

Building rapport with your stakeholders will go a long way toward developing productive working relationships. Even though you're engaging in a professional context, it will help you work well with colleagues if you have a sense of each other as people, behind the masks that we all need to function in the workplace. That will build trust among the team, as well as helping you understand what motivates each of them.

The key social skills to develop are how to be relatable, how to build mutual respect, how to practice empathy, and how to encourage vulnerability.

- To become *relatable*, seek out common interests or experience to build connections with your stakeholders. Finding similarities with a diverse set of stakeholders will leverage affinity bias in a positive way. If you can't find existing connections, make some new ones.
- We all have an innate desire to be *respected*, which means being treated with due regard for our feelings, wishes, and rights. To respect others, you must recognize that people likely have good reasons for their behavior and opinions, even if they are puzzling to you.
- *Empathy* builds rapport because we show people we understand them. Being curious and making people feel heard are key ways to demonstrate empathy. You can do this by asking open-ended questions, using active listening, mirroring, and summarizing.
- *Vulnerability* is openly acknowledging what's really going on, being willing to take on risk, and accepting the possibility of failure. Create psychological safety for your stakeholders by demonstrating that they will not be reprimanded for stating an unpopular opinion or making a mistake.

Take their point of view and solve their problems, just like you do with customers

Chapter 4

Trust

Building rapport with stakeholders takes effort, but building a good working relationship is not the only thing you need to be successful. You also need to prove to your stakeholders that you are credible, by demonstrating expertise and confidence, so they will trust your decisions instead of second-guessing you. And you need to show them that you are reliable, by demonstrating ownership and dependability, so they will trust your process instead of micromanaging you (Figure 4-1).

Figure 4-1. The elements of trust

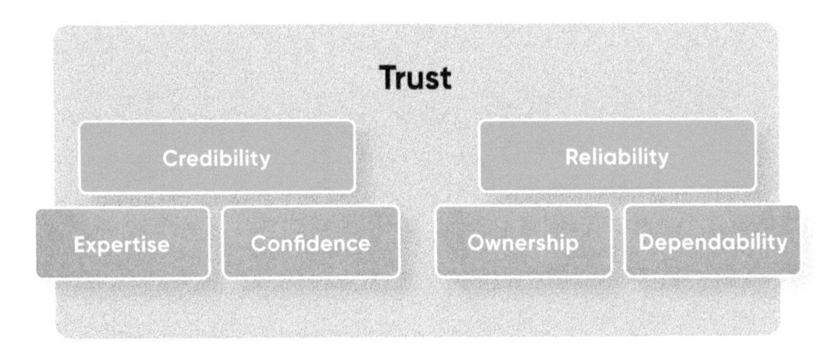

In this chapter, Irie will continue building trust with her stakeholders, and you will learn some key skills:

- Demonstrate expertise without sounding like a know-it-all.
- Show confidence, be prepared when you speak with your stakeholders, and provide transparency into your processes.
- Take ownership of your product, identify risks early and propose mitigations, and own both your successes and your failures.
- Become a dependable resource, be responsive and deliver on your promises, and set the right expectations up front.

We start this chapter as Irie has just received an opening from Sparks that he might accept her help with his board presentation.

4.1 Irie Needs to Show Her Expertise

Irie is chatting with José, the director of design, over lunch. "I think Sparks has given me an opening to help him with a board presentation on the AI ideas we've outlined," she says. "Have you worked with him before on something like this?"

"Not really," says José. "Sparks usually holds these board presentations close to the vest. But I'm not sure he'd trust me to help anyway."

"Why is that?" asks Irie.

"I feel like he doesn't believe in my role here. He's always questioning the value of user research."

"I'm sorry, that stinks," says Irie, demonstrating some empathy for José's situation. She pauses to see if he will keep talking. He does.

"I think I haven't done a good enough job in proving myself here," continues José. "My team has done some great work, and we've identified some real product winners through our research, but somehow the credit always goes elsewhere. Once, Sparks proclaimed in a company all-hands that he discovered that customers were confused about the onboarding process and told the engineering team to fix it, when actually it was my team that made the discovery. He touted himself as the hero!"

"Wow," says Irie. "That's tough. I feel like the same might be true for the product management team. People here don't seem to know what we do. Even the fact that we both report to the technology department means that some people might think we're just an extension of the engineering team. Maybe we can work together to make it a little clearer what we do?"

"That would be great," says José. "But how do we do that?"

"Good question," says Irie. "I think this might warrant a call to Darius, my former boss."

"Do you often call former bosses when you need advice?" asks José, with a puzzled expression.

"No, not usually," Irie says with a smile. "Darius is an advisor now. He gives advice to product leaders. He told me I could always call him with quick questions."

"I'm not sure that this is a quick question," says José.

"I guess we'll see!"

Irie texts Darius, and sets up a time for later that afternoon.

When she and José join him on a video call, Irie makes introductions and explains their concern. She finishes by saying, "We are both feeling like people don't respect what we do."

"It can take a while to build up that kind of trust," Darius says. "Our company had a strong engineering culture when I joined and it took me a while to establish the role of product management before we really found our footing. Design took even longer, actually."

"What did you do to get over that hump?" asks José.

"I tried explaining how it worked at places I'd been before but that got us nowhere," Darius shares. "Explaining my point of view and my team's needs just bounced off of people who'd been there for years and had established ways of working."

"But you eventually got there," says Irie. "When I worked for you, product and design were equal partners with engineering. And you were a member of the executive team alongside the CTO. I never had to explain what we did or that our team owned the roadmap."

"That shift started when I started listening instead of talking," Darius explains. "I needed to demonstrate that I understood how things worked, that I understood what problems the other teams had, and that my team—including you, Irie—could contribute. It really started by focusing on their needs," Darius adds. "You can try to speak their language, demonstrate relevant experience, and even just summarize what they've said to show you understand."

"I think I see where you're going with this," says Irie.

"And if people really don't understand what product and design do," says Darius, "it might be worth explaining, but I'd still focus on their problems and show them how you can help. They'll adjust their process and their assumptions when you demonstrate your value. It's showing rather than telling."

"Take their point of view and solve their problems," says José, nodding, "just like we do with product users." ■

Expertise

The best approach to demonstrate expertise is "show, don't tell." Counter-intuitively, an unsubstantiated assertion about expertise, like "I know what I'm talking about" or "just trust me," makes people trust your expertise less instead of more. The better approach is to show them what you know, and let them draw their own conclusions about whether you're an expert.

Speak their language

Using the right jargon for your stakeholder shows you've done your homework. When you use too much of *your* jargon, they will doubt you because you can't even communicate in "plain English." But when you use *their* jargon, it demonstrates knowledge and familiarity with their world, which leads them to trust your expertise. For example, if you're in the beverage industry, you might say "soda" when speaking with a stakeholder from the northeast US, but "pop" when speaking with a stakeholder from the midwest, and "Coke" with a stakeholder in the south, utilizing familiar words with each audience.

When entering a new industry, starting at a new company, or working in a specialized field, preparation is your friend. If you hear a stakeholder use a term you don't recognize, write it down and look it up. You can even build a glossary of definitions and acronyms and study it to help you converse with them.* If a list of key terms and acronyms isn't a part of the new-hire onboarding process, maybe you can create one.

Melissa once worked on a task management product for IV nurses.† Her stakeholders were former doctors who had started a company to solve a familiar problem. It wasn't familiar to Melissa so she researched terms she might encounter and created a three-page glossary. When she used these terms with the stakeholders, she could see their eyebrows raise and their shoulders relax, with a confidence that she was familiar with their world. This made them more comfortable talking about the problems they were seeing, because they didn't have to constantly pause to figure out how to explain it to a layperson.

---

* We have both done that multiple times in our careers.

† IV nurses specialize in putting in, taking out, and managing IVs in hospital patients, usually at larger hospitals. Because they manage IVs all day long, IV nurses become adept at doing them well and noticing problems, which can reduce the chance of infection.

It's okay to ask for definitions when you are new in a role, or being introduced to a new person, or learning a new product area. But once you've been in a role for some time, it can feel awkward or make you seem inexperienced if you constantly ask your stakeholders what common terms mean. The trick is to find a trusted peer to answer your questions, like a fellow PM or an engineer or designer. This will help you quickly build up your knowledge without revealing those gaps to your stakeholders (and the gaps can be quickly closed anyway with the help of your peers).

If you're having trouble with the jargon, or you're in a conversation unexpectedly without time to prepare, you can summarize to show your understanding. Just as summarizing can demonstrate empathy (see Chapter 3), it can also be used to show expertise, by showing that you understand their situation well enough to sum it up in a few words. If you get the summary wrong, that's still helpful, because your stakeholder can correct you and expand on their thoughts, giving you an even deeper understanding.

Share relevant experience

Giving specific examples from your past experience will also help you show your expertise. When stakeholders realize you've actually done this before, and with good outcomes, they will be more likely to trust your advice for their situation.

Understanding the perspectives and motivations of your stakeholders will help here as well. By combining what has worked in the past with an acknowledgment of your stakeholder's fears and concerns, you can show that you are capable of providing a good solution to the problem.

For example, Melissa was working with a sales director who was uncomfortable having the product managers talk directly with customers. Melissa shared how in her past roles, the close contact between the product management team and customers was an asset to the sales team, because customers saw their closer relationship with the product managers as a benefit of choosing their company over the competitor.

Melissa shared with the sales stakeholder some structured ways of working with customers that she used in the past, such as customer advisory boards.* She also mentioned ad hoc customer interactions, where sales team members could select which customers would be invited for interviews. Citing tools and techniques that worked in the past, while recognizing the desire for the sales team to control customer interaction, convinced the sales director to give it a try.

When using the "relevant experience" approach, there is a delicate balance between giving helpful advice and sounding pushy. Sharing your relevant experience, but letting the stakeholder choose whether to follow it, gives them a bit of control while also conveying that your suggestion is based on your experience. You can show deference using phrases like "I'm not sure if it would work here, but in the past I've..." or "We don't have to do it exactly the same way, but last time I ran into this situation I fixed it by..." Humility frames your advice as helpful, rather than all-knowing.

Provide targeted education

If all else fails, you might need to provide some context on the role of product management, and why it's valuable. You could do this through one-on-one conversations, or a group "Lunch and Learn" format, depending on the appetite of your stakeholders to learn more about your team.

Helping your stakeholders understand the benefits you can provide is essentially internal marketing for your role and your team. Start by understanding your audience and their problems to better articulate the benefits that product management can bring.

Because product managers usually have a high-level vantage point across many disciplines, plus a deep understanding of customers, they are uniquely positioned to be useful to many people across the organization. Here are some examples of stakeholder concerns and how you might explain the role of product management in fixing their problems (Figure 4-2).

---

* A customer advisory board (CAB) is a group of customers who are recruited to act as advisors to the Product Team, and is primarily utilized with B2B products.

Figure 4-2. Stakeholder concerns and how product management can help

Stakeholder type	Stakeholder problem	How product management can help
Customer support	I want shorter lead times on delivering on customer requests.	Product managers can talk with the customer to define the problem and work with engineering to determine a solution and delivery timeframe.
Sales	I want to be able to promise a solution to the customers' problems.	Product managers can synthesize information from various internal teams and outside sources to determine which problems, when solved, are likely to enable the most sales.
Marketing	I want to know which features we will ship six months in advance so I can plan my marketing activities.	Product managers can help you understand the goals, needs, and wants of customers in order to improve messaging and communication. Focusing on customer problems and benefits is usually more effective than touting features.
Operations*	I want the solution to be usable by internal users and easy to launch.	Product managers think about problems holistically, incorporating features, user experience, process, launch, field training, safety, scalability, cost, compliance, risks and mitigations, and how to measure success.
Security, compliance, legal	I want to ensure our products comply with laws and regulations.	Product managers can help you understand the technical options and the cost of those options, to help weigh the risk of noncompliance with the cost to comply.
Executives	I want to make sure the products we launch will sell well and have a good return on investment.	Product managers can do the customer research and product discovery before committing to solving a particular problem, to ensure that it is worthwhile. We can test proposed solutions before engineering gets started to make sure customers will find them useful rather than investing money and resources into creating solutions that don't end up selling.

* Different industries use different operations functions. Most companies have a customer service team, an online retailer might operate warehouses, a SaaS company might operate data centers, and a solar panel company might have field technicians. Many operations teams depend on internal product teams to support them with robust products they can deploy and use easily.

Irie Demonstrates Her Skills

Following their conversation with Darius, Irie and José, along with Yacob, the engineering director, are sharing ideas for describing how they can help different stakeholders. The trio are sitting in a huddle space toward the back of the office, by a set of windows.

"We're like a three-legged stool: product, design, and engineering," Irie says. "Our stakeholders need to know how we operate and how to interact with us."

"Some of them get it," chimes in José. "Like Ella, the CRO. Her customer support team is actually really great at introducing us to customers and letting us lead the conversation."

Yacob interrupts. "But her sales team is terrible. They come directly to engineering all the time with customer requests that aren't related to any of our quarterly goals."

"We should really be funneling all requests for features or bug fixes through my team," Irie says. "Your teams shouldn't have to deal with this stuff." Irie explains that she's working on creating a single queue for requests to make this easier and the other two agree quickly with evident relief.

They develop a list of typical problems by stakeholder role and discuss how the product management team can help with each. "If people see how we can help them," says Irie, "maybe these conversations will go a little easier."

After this discussion, Irie walks back to her desk. A tiny woman in a brightly colored dress is waiting there for her, looking impatient.

"There you are!" exclaims the woman. "I have been looking all over the building for you!" She is staring challengingly at Irie, who responds by pointing at herself quizzically. "Yes, you! You are the new product manager for the app, no?" Irie nods. "Well, my customer is not happy and I need you to help me get something on the roadmap."

The woman is Arianna, in sales. She says that she has received an ultimatum from one of the company's largest corporate accounts, RightBank, which enrolls their people automatically in the Helthex app as part of their employee benefits. "They require us to have user access management or they will not renew. I have the contract ready to go, but legal tells me I need you to sign off!"

Irie takes a moment to process this outburst. She formulates some questions in her head to help her understand the request better. She also thinks that the new product request intake process can't come too soon.

"Well?" asks Arianna impatiently.

"Let me make sure I understand the request," says Irie. "One of our biggest accounts won't sign the renewal contract unless we provide user access management, is that right?"

"Yes, but I already have it in the contract, so I don't understand why I need you to sign off on it," clarifies Arianna.

"Okay, so the problem is that you believe we should be able to deliver this, and you just need to make sure it will be ready by the time the new contract starts?"

"Hmm," says Arianna. "It would help to be absolutely sure it will be ready on time."

"Right," says Irie, doubling down on what seems to resonate with Arianna. "We don't want to disappoint the customer by promising something and then not delivering it."

"I certainly don't want to disappoint them, but why can we not just deliver it?" asks Arianna.

"I will check into it. When does the new contract start?" asks Irie.

"Not for another three months," admits Arianna.

"Okay, here's my proposal. I'll check with engineering and design to see what it would take to deliver a user access management feature. Hopefully, we can deliver it in three months, but I can't guarantee that on the spot here without talking to them."

"So your job is to gather the requests from sales, ask engineering and design for estimates, and tell us when they will be done?" asks Arianna.

Irie perceives her role through Arianna's eyes as little more than scheduling. She thinks of the conversations she just had with José and Yacob and begins to formulate a reply, when Arianna continues, asking, "Why don't I just ask them myself?"

"Deciding what goes on the product roadmap is part of my job," Irie begins, "but that's really just the end of the process. It begins with understanding the market—all of the customers and prospects, not just the ones we are talking to right now. We gather information from every available source, including the sales team, of course. We look at the competition, we look at what's feasible, we develop a strategy, and we try to determine what will maximize results for customers and for the business. At the end of this process, that's what goes on the roadmap."

"So if user access management does not fit your strategy, you will not do it?" asks Arianna.

"That is a possibility," Irie replies calmly, "but that seems unlikely here, since user access management is typically included with products aimed at larger companies. Assuming other customers also need it, we need to look at the rest of our priorities. I know we have some other things in the works for RightBank and other customers, like enhancements to the admin console. We'll need to evaluate our initiatives against one another so we don't deliver for one customer while breaking our promise to another."

"I don't like the idea of pitting one customer against another, or one sales person against another," says Arianna. "But this must be the top priority. They are one of our largest customers!"

"We only have so many engineers," explains Irie, "so we focus on the things that will deliver the most value across all customers, not just one." Arianna's eyes widen at this, so Irie asks, "Can you find out how many corporate customers might benefit from user access management? If this problem is widespread we could be looking at a winner here...assuming it ends up being easy to build."

"I can do that," says Arianna. "I believe there are many customers who would value this capability." ■

4.2 Irie Does Her Research

After speaking with Arianna, Irie asks Yacob about the user access management feature.

"We've looked into it," Yacob says. "It's come up a few times with corporate customers. They have a lot of users, so they want tools to add and remove users and change permissions. But it's not as simple as it sounds. Our platform doesn't have the concept of different types of permissions, because we built it with individual B2C customers in mind. It's a lot of work to add that to the platform, plus we'd need to do the research to figure out what different types of permissions we'd even need to support. What we're working on now is the ability to bulk-upload user lists, which I think is a reasonable workaround, but some corporate customers are still not happy."

"I'm surprised we don't already have user access management, but I guess it makes sense if we wanted to move quickly and we weren't planning on corporate customers. How much work is it to add different types of permissions to the platform?" asks Irie.

"It's significant. It would probably take six months, because we only really have two engineers who can work on platform changes like that, and that's not including the other work they're already doing. We haven't invested in more platform engineers because we have been focusing on improving the UI. We could try sourcing some contractors, but it would take some time to find them and onboard them, so I still don't think that would put us inside the three-month timeframe that Arianna is asking for."

"It sounds like there's not much we can do to speed it up to three months, but six months might be possible," says Irie. Yacob nods his head. "So what do we do now with Arianna's contract renewal?" Irie asks. "I wonder if the customer would be willing to sign now with a delay in delivery? Not sure if that would work."

"Dunno," says Yacob. "I don't even know if I can commit to six months right now, because there are a few other platform projects I already know about that seem to have equal priority to this one."

"Plus the AI work that Sparks wants will probably also need the platform engineers," says Irie. "We should probably have a resourcing conversation, because your engineers are not interchangeable."

"Definitely," says Yacob.

"I guess I'll have to be the bearer of bad news to Arianna," sighs Irie.

Sri is working from home today because one of his kids is home sick, so Irie tries to find a quiet spot to dial into her regular one-on-one with him. She sneaks into Liz's office because it's empty. She feels weird about sitting at Liz's desk, so she chooses the small couch in the corner of the room. Liz is the CEO, after all.

"I'm not sure what to do, because most of our corporate customers want user access management, but we don't have the resources to deliver it right now, especially with the upcoming AI work," Irie says. "I'm waiting to hear back from Arianna on the total opportunity. But I just spoke with Yacob and he says it will take six months to get it done, and we have all these other commitments too."

"You need to be honest with Arianna," says Sri. "If you can't deliver, you need to tell her that now. It's a good idea to set the correct expectations from the beginning. It's better for people to be disappointed now rather than getting their hopes up and being disappointed later."

"True, but I'm trying to build trust with Arianna, and this could make her think I'm not capable of doing my job if I can't figure out a way to get this feature done."

"So just be transparent with her," says Sri. "Tell her what's going on and why you can't do what she's asking."

Irie thinks about this for a moment, then replies, "That's right, I think. But it's also my job to ensure we have good answers for our customers. I'll talk to her about getting closer to this customer."

Sri approves. "Sounds like a learning opportunity."

"By the way," adds Irie, "I saw Sparks in the hallway earlier today and followed up with him about the board presentation I had offered to help with. He told me he didn't think I show up as 'confident' in the meetings he's been in with me, so he didn't want me to talk to the board. What's that about?"

"Sparks uses that one all the time. 'Confidence' is his go-to excuse for blocking people from getting visibility with his buddies on the board. Particularly young women, unfortunately. He's not very self-aware when it comes to his biases."

Hearing this from Sri makes her feel a bit better about the interaction with Sparks. She decides not to take this snub from him personally. "I told him I'd send him some materials for the board presentation and he can decide whether or not to use them."

"Good idea," says Sri, his trademark smile returning. "That seems like a pretty confident move to me."

After their meeting, Irie feels like she's out of ideas on how to solve Arianna's problem with RightBank. She considers just laying all her cards on the table and telling Arianna what she's up against. *Arianna will definitely be disappointed*, Irie thinks, *but maybe I can show her that I am still good at my job by showing her how thoroughly I've looked into it, and also by offering alternatives for helping her customer.*

Irie thinks about how both being prepared and being transparent can be forms of confidence. ■

Confidence

In the strictest sense of the word, confidence is a belief in one's own abilities. It's difficult for other people to believe that you have the skills required to do the job if you don't believe in yourself. In product management, confidence is especially important when communicating decisions and the reasons for those decisions. Delivering decisions with confidence helps stakeholders trust that you've thought it through, and that it's a good solution.

People with a quieter leadership style (especially women)* have been unfairly accused of lacking confidence. Being confident doesn't necessarily mean being loud or assertive—or even an extrovert. It means being sure of yourself, knowing that you've done the work necessary, and believing that you can achieve positive results. Sometimes this takes the form of detailed quantitative analysis and a narrative that effectively tells your product strategy story to your stakeholders. Sometimes it means that you barely have to say anything in an alignment meeting because you've already pre-aligned with each party individually before the meeting.

Regardless of how you show it, you need to make sure your stakeholders understand that you believe in your decisions and recommendations. Confidence has two basic components: preparedness and transparency.

Preparedness

Being prepared is a good way to show confidence, and thus credibility, because you have done the work to "have all the answers." Think of a situation in which you've had to present certain information, like a product strategy, roadmap, or decision. Many of us have been in such meetings that didn't go well because we were hit by unexpected questions, and we didn't have good answers. That can definitely throw you for a loop and dash your confidence! You can prepare by predicting the questions, which can be done by yourself, with a partner, or with a small group.

If you've done your homework to understand your stakeholders, as we discussed in Chapter 2, you should have a pretty good idea of how they think and what motivates them. You can use this information to predict questions they might ask. For example, you'd expect someone in finance to ask

* For a full analysis of the confidence gap between men and women, we recommend *The Confidence Code* (Harper Business, 2014) by Katty Kay and Claire Shipman.

about the monetary impact of a decision. You'd expect someone in sales to ask about timelines and when features will be ready to sell. So prepare the answers to those questions before you meet with the stakeholder, and even send the answers in advance, which will give you more confidence in your conversation and will build trust with your stakeholder.

Another way to prepare for a big meeting is to make sure you're in a good state of mind: get enough sleep, eat well, and exercise. You can even wear your favorite outfit to give yourself a little boost of confidence.

Transparency

Even if you don't yet have all the answers, you can show that you've done your homework by listing the things you are still working to answer. Here are some phrases you might use to show confidence through transparency:

- "I thought of that, but we don't have all the data we need yet to provide that answer."
- "When we looked into it, we discovered that there's a dependency on another team, so I will have to get back to you on the timeline estimate."
- "I have a meeting set up with Joe next week to discuss that very question, so I should have the answer after that meeting."
- "That's a great point, I hadn't thought of that. I will go chase that down."
- "That's a really good question. I'll look into it and get back to you. Who's the best person on your team for me to connect with?"

When you have to make a tough decision, being transparent about how you got to that conclusion helps your stakeholder have confidence in your ability to do your job well, and improves your credibility. Even better, you can share how the stakeholder helped you arrive at the best possible outcome.

You can demonstrate transparency in several ways (Figure 4-3).

Figure 4-3. Transparency dimensions

Transparency dimension	Key idea	Explanation
Decision making/ exploring options	What's the plan?	Share not only the great ideas you have, but also the ideas you've decided are not going to work.
Process/methodology	How are we executing the plan?	Include your stakeholders in the process and help them understand the reasons behind each step.
Risks/challenges	What is going or could go wrong?	Call out risks and challenges early and make mitigation plans for them.
Status/learnings	What's happening now?	Share frequent updates for high-profile initiatives. Call out anything that changed recently and how that affects your plan.
Personal/opinion	How do we feel about it?	Share how you're feeling about key decisions. Maybe your stakeholder feels the same way and you can build a connection.

Transparency isn't sharing everything you know. Share only relevant information, especially if someone has told you something in confidence. For example, if you know that an engineer on your team is thinking of quitting, you don't have to share that information with your stakeholder. Instead, you could point out a general risk of attrition on the team. Calling out that risk is also a form of transparency.

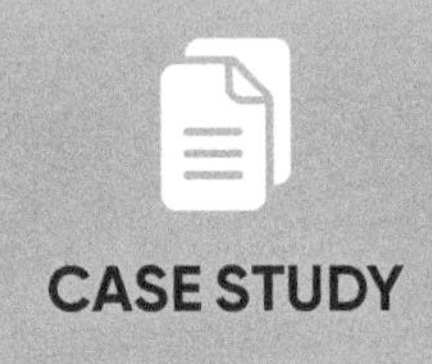

Using Transparency Dimensions

Melissa worked at a company that had scaled their product for a number of years before realizing their technical debt had gotten out of control. The engineering leadership team finally decided to take action when there was so much work just to keep the current product running that they literally had no time to build new features.

This presented a tricky situation for product managers working with their business stakeholders. PMs had to explain to stakeholders not only why they were unable to solve critical business and customer problems, but also why they would need an additional 6 to 12 months to dig out of the hole. Melissa and the technology leadership team created a narrative for stakeholders that involved transparency in a number of these categories:

Challenge: The product breaks constantly and takes a long time to fix. This is due to a number of factors, such as inefficient systems, lack of consistent monitoring, unsupported components, and a patchwork of "quick fixes" over the years that are no longer holding together. Because of this technical debt, we are unable to solve new problems without aggravating the existing stability problems.

Personal: Our technical team has made some less-than-optimal decisions in the past, in order to support business objectives. We have now learned that there is a limit to how often we can use the "easy" way to implement a new feature. We don't like having to say "no" to new features right now, but we don't really have another choice.

Decision: Our effectiveness as a team is grinding to a halt. We need to dedicate significant engineering resources to identify the biggest opportunities for improvement and to fix those key problems. Without doing this work, the risk is too high that adding new features will make the product unmanageably unstable for our customers.

Process: We will focus primarily on reducing technical debt while including selected features. All new features will be built "the right way," focusing on the most relevant technical debt as we go. We need you, our stakeholders, to help us ruthlessly prioritize our backlog. While the engineers are doing the technical work, the product managers and designers will do customer discovery work to identify and validate key problem areas, so that we will be ready to pick up speed when the stability work is complete.

Status: We will keep you updated on our progress as we get our product back to a stable state. We are working now to estimate how long that will take, and we will likely revise our estimate as we get into the work.

Providing transparency into the thought process and the plan forward convinced stakeholders that the team was confident in their assessment of the situation and confident in the proposed solution. Because the team was confident, the stakeholders trusted them that prioritizing the technical debt was the right decision.

Irie Shows Confidence

Irie is still wondering about the cost of the user access management work versus the value. She wonders if her new acquaintance from Finance, Justin, might have some perspective.

Dropping by his cube, Irie discovers Justin is absent. She glances around and catches sight of him in the nearby kitchen. She finds him there staring at an array of snacks.

"Why fool yourself into thinking that a protein bar labeled 'cookies and cream' is healthy?" Irie says. "If you're craving something sweet, go directly for the Skittles, I say." Justin can't think of a counterargument, so they both opt for candy.

After some chatting, Irie decides to bring up the topic that's on her mind. "I heard you talking about expenses versus revenue earlier with Pria. Do you know much about our margins?"

"Actually, that's my job," Justin explains. "I review all contracts with customers and partners. We need to watch how much we discount with big companies. We've been burned before with unprofitable deals."

"I thought we sold direct to consumers," Irie says, before tossing a few Skittles in her mouth.

"Mostly, but we have these few really big deals with corporate customers," Justin explains, unwrapping his Snickers bar.

"How much of our revenue comes from corporate customers?" Irie asks.

"It's like 10%, but it's growing. The problem is it's less profitable. We kinda 'break even' across those deals right now."

"You said you review all of these contracts," Irie says. "Do you happen to know how many of them are asking for user access management?"

"Oh, that always comes up with corporate deals," Justin says. "I'm told we can't do it, so we always say no. There is tearing of hair and rending of garments, but eventually, we negotiate a lower price and they sign."

"Hence your job," Irie observes.

"Check," says Justin making a swishing gesture with his free hand.

Thinking for a moment, Irie asks, "In these corporate deals, are there a lot of features we have to negotiate around?"

Justin's eyebrows climb to the top of his forehead. "Are you kidding?" he asks. "Every deal has a list the length of the Mississippi River. And they're all different! We try to say no, but there are always a few deal killers we need to put on the roadmap."

"So, you're saying that the corporate business is relatively small, that it's unprofitable, and that it requires us to do a bunch of feature work to close deals. Is that right?" asks Irie. Justin spreads his hands in a helpless gesture. "Why are we doing these deals?" Irie asks.

Hands still in the air, Justin says, "I don't make strategy. I'm just keeping us out of trouble."

Irie meets with Arianna the next day to talk about what they've each learned about user access management. She is dreading the meeting, because she doesn't like sharing bad news, but she knows it's important.

"It turns out you were right," begins Arianna excitedly. "Many of our customers want user access management. So this means we're definitely doing it, right?"

"Well…" begins Irie.

"Uh-oh," says Arianna. "I don't like the sound of that 'well.'"

"The bad news is that the engineering team has looked into it and they've determined that it's going to be more like six months, not three months, to do the work." Irie explains the specifics of what Yacob told her about the technology. "And even if we commit to it, there's still a risk it won't be done in time, because there are only two engineers who can do this type of work. Plus, remember, we still have to evaluate this against our other priorities."

"So what am I supposed to do about this contract?" demands Arianna.

"I know, it's really not a great situation. I feel bad about not being able to deliver what you want, because my job, just like yours, is to make the customer happy."

"I don't see how any of this is making customers happy!" says Arianna. "I should have just had the contract signed without talking to you. Then you would have been forced to deliver!"

"But if it actually takes six months and we tell them three, that would have led to broken promises and unhappy customers," Irie reminds her.

Clearly frustrated, Arianna looks away.

"Not all is lost," says Irie, reassuringly. "I think we can find something else that will deliver value to the customer. It just might not be exactly what they asked for. How would you feel about me talking with them? You can be on the call too, of course."

"Oh, I don't know about that," says Arianna, crossing her arms in front of her. "Contract renewals are a delicate situation."

"I understand that," says Irie. "I don't want to mess up the work you've done with them. But maybe I can help. Customers are often happy to talk with me because I can actually make changes happen in the product."

"As far as they know, I can make changes happen in the product too!" says Arianna. "But I see what you mean."

Irie changes her tactic a bit. "They're also really happy when the product management team listens to them and asks for their input. So for example, we can ask how they would prioritize this feature against others on our list, or how urgent this request is compared to something else. That gets them thinking about whether they really need this to sign the contract. And we might uncover different needs that we can solve right away. That builds trust that we can deliver, even if we can't deliver their original request."

Arianna looks appraisingly at Irie.

"And," Irie adds, "I can be the bad guy and say 'no' so you don't have to."

"That would help," says Arianna, sounding a bit relieved. "It would allow me to keep my relationship with the customer positive."

"Exactly," says Irie. "This is a joint effort. They are my customer too." ■

4.3 Irie Demonstrates Ownership

First thing Monday morning, Irie is making coffee in the kitchen near her desk, when she notices Sparks walking down the hallway toward her.

"I saw the material you sent me for the board presentation," Sparks says to her without preamble. "It's actually great stuff."

"You sound surprised," says Irie wryly.

Overlooking her sarcasm, Sparks continues. "It lays out some good ideas for how we can use AI in the product, and it has research to back it up. How did you get this done so quickly?"

"Well, it wasn't just me. Eitan and Christina on my team, along with Divya in data science and José in design, have been thinking through the AI topic for a while now. I just helped them bring all the research together into a format that would be good for the board."

"Well, I appreciate it," says Sparks. His voice drops to a whisper. "I think the board will be happy that we're doing all this work."

"No worries," says Irie. "I'm glad you liked it." Irie pauses to think, then asks for clarification. "'Doing all this work' to put together the ideas, right? Because it's too early to say which ideas will work for customers, what's feasible, or when we can deliver. Also, we need to think through how the AI work fits into the rest of our strategy and priorities."

"I don't understand why we need all of that," replies Sparks. "We have the roadmap that you laid out, now we just have to execute."

"What I gave you was not a roadmap," Irie says emphatically. "It was just a list of preliminary ideas. I was hoping to run the ideas by the board to see if they have any initial feedback. We can't make any promises yet." Irie then realizes that Sparks hasn't been paying attention to any of their conversations about the product development process. "I can walk you through our process again, if it's helpful."

"That's not necessary," says Sparks. "I get how it works. I didn't get here yesterday."

Irie sighs. She knows she's not going to win this argument today. "Maybe I should present these ideas to the board or answer any questions. I want to make sure they know these are not promises. A roadmap should lay out our strategy. Developing features is only one piece. We have a lot of work to do before we have a roadmap."

"No, I've got it," says Sparks, perusing the snacks and grabbing a bag of chips. Irie thinks that's an interesting choice for breakfast. Sparks adds, "But I really appreciate the deck you sent and all the work that went into it." As he walks away he says, "We'll probably end up getting outside help on most of it, so don't worry if the engineering team can't handle it right now."

Before she can ask any follow-up questions, Sparks is gone. Irie sighs again as she stares into her coffee.

Later, Irie is pacing outside Sri's office, trying to decide how to tell him what happened with Sparks.

Sri spots her from inside his office, comes to the door, and says, "You look upset. Shall we take a walk?"

As they walk through the business district outside their office, Irie makes her confession. "Sri, I think I messed up."

"How so?" he asks. Sri eyes the coffee shops as they pass by, but Irie is too absorbed in the problem she's trying to solve to notice.

"Remember the AI ideas I sent to Sparks? We agreed it was a 'confident move,' but now I'm really worried. He seems to think I was sharing a committed roadmap, instead of just some ideas. I think he's planning to tell the board we will build all of that in the next few quarters. I tried to explain how much more we have to do, but he just walked away. I don't know what to do!"

To Irie's surprise and embarrassment, Sri laughs.

"What is so funny?" Irie demands. "Sparks is about to commit our team—your team—to build a bunch of untested ideas in an undoable amount of time."

Still smiling, Sri says, "Welcome to Helthex!"

"Sri," Irie says, "we will fail. People will quit. And Sparks will blame it on us!"

"No, no, you're right," Sri says, sobering a bit. "But I'm laughing because this is Sparks's thing. He gets hot on an idea, sets unrealistic goals, gets everyone to run around pursuing them, and then in a few months he gets hot on something else. The team knows it, even the board knows it."

"But this is not the way we should be doing things," Irie says. "We'll never make progress on anything if we switch to something else every few months."

"And this is why I hired you," Sri says.

"I know how to make a proper roadmap," Irie says. "We need a vision, we need objectives, we need a target customer, we need a strategy. We have Liz's vision, but we don't have any of those other things!"

"How have you done those things in the past?" Sri asks.

"The executive team worked on them together," Irie explains.

"Who brought them together for that work?" Sri asks.

"At my last company it was Dariu—" Irie begins and then stops. Sri waits for her to continue. Eventually, she says, "So you're saying that this is my job as head of product."

Sri's smile broadens. "Liz always kept him in check but she can't do that from the sidelines."

"I don't have that kind of clout," Irie says. "You're on the executive team. You've been here, established credibility..." she trails off as Sri shakes his head.

"I can't do this," he says. "I've tried but I just don't have what it takes. You know I hate running meetings, the politics, the personalities. You have the product leadership skills we've been missing. There is no one else."

Irie lets out a long breath. "Sparks doesn't really want me to do that job," she says. "But it has to be done or we're going nowhere. I think I know what to do. I'm guessing that if we bring the whole exec team together, we can come to some sort of alignment. But I can't do it alone."

"I didn't mean to say you'd be on your own," Sri says. "Like you said, it will have to be a team effort. That includes me. I'll support you 100%, but I need you to take the lead."

Irie mulls over their conversation as she and Sri walk back to the office. She begins thinking about how she might bring people together and what they would need for inputs to have a solid statement of strategy and direction.

As they arrive at the building, her mind snaps back to the upcoming board meeting. "You'll be at the board meeting, right?" she asks Sri. "I still want to head off making any premature commitments."

"Actually, I have an MRI for my knee tomorrow," says Sri, "so I asked Divya to cover the staffing plan at the meeting."

Irie is momentarily caught between the sudden realization that the board meeting is tomorrow and this news about Sri. She settles on concern for her mentor. "What's going on with your knee?" she asks.

Sri explains that he's had increasing pain since overdoing it playing soccer a few weeks back.

"Oh, and I made you walk all around town!" Irie exclaims.

"That was my idea, actually," he reminds her. "It stiffens up if I sit too much and a short walk helps loosen it up."

"Oh, that's good," says Irie. "I don't feel so bad now. But I didn't realize the board meeting was tomorrow! There's no time to get in front of this."

"I assume this roadmapping process you described usually takes weeks," says Sri.

"Yes," Irie confirms, "or months."

"So with the board tomorrow we really just need to buy some time," he says. "Divya knows how uncertain all of these ideas are, but could you outline the biggest risks of overpromising? We can brief Divya and she can try to contain things tomorrow. She's seen this sort of thing with Sparks before." ■

Ownership

You might think you can be successful at your job by simply executing on all the tasks assigned to you by your manager. But the best employees exhibit real ownership—they consider factors beyond their explicit responsibilities to execute more seamlessly and reduce risk. Ownership for product managers is knowing about, caring about, and taking responsibility for everything that goes into making a great product.

Sometimes ownership is as simple as making your own plan instead of waiting for someone to tell you what to do. Bruce's first raise came from his first boss, Bill. Bruce painted houses during college, and Bill told him he didn't want to have to stop work to direct Bruce to the next assignment every time he finished a task. He said he would pay 10% more per hour if Bruce could figure out for himself what the next task should be and simply tell Bill what he planned to do. This lesson in ownership and initiative has stuck with Bruce throughout his career.*

Beyond basic initiative-taking, product managers are responsible for ensuring successful product outcomes, which requires participation from a number of different types of stakeholders. Ownership means both identifying these key stakeholders, and getting them aligned on participating in your plan. Sometimes you will need help from unexpected teams, like field training or procurement or contracts. Ownership means taking the initiative to understand all the functions in your company and utilizing them effectively.

Ownership often involves taking on unexpected challenges and getting creative to ensure the desired outcome is delivered, regardless of the obstacles along the way. To overcome these roadblocks, you need not only to involve your stakeholders, but also to proactively assess risks that might arise. To quote the Whether Man from Norton Juster's *The Phantom Tollbooth* (Random House, 1961), "Expect everything, I always say, and the unexpected never happens."

* Thanks, Bill Maxwell. You also taught me how to paint windows without masking tape, and how to run a small business!

Levels of ownership

As with any skill, ownership expectations increase as you get to a more senior level in your career (Figure 4-4). These expectations are set by your manager, but are also shared by your stakeholders.

Figure 4-4. Progressive levels of ownership

Level 1: Execute well

Even at a junior level, you should take pride in your work, and create high-quality deliverables. Take advantage of learning and mentorship opportunities where you can find them, to ensure that you have the best foundation to do a great job on your responsibilities.

This doesn't mean that you have to do everything on your own. You can ask for feedback from peers, direct reports, your manager, or a trusted advisor. In fact, asking for help shows that you care more about the quality of your work than you do about your own ego. And getting feedback on your ideas usually makes them better. If you manage a team, you can utilize the various skills on your team to create the best possible outcomes.

Level 2: Involve stakeholders

This whole book is about bringing various types of stakeholders into your process, so by now you should understand the importance of working with people and teams outside your own to drive product success. Stakeholder alignment is not a nice-to-have, it's a must-have, and it demonstrates ownership. Remember, software delivery alone does not make a successful product.

Stakeholders (e.g., marketing and sales) and other Product Teams depend on you—and your product's success depends on them. Identify those dependencies early so that you can manage the implications of your decisions beyond your Product Team, and beyond your immediate part of the product.

Level 3: Be proactive

Going one step further, you need to do all of this without specifically being asked. Proactively reach out to the sales folks before they come to you with requests. Build relationships with others in your company so that you know about potential conflicts or dependencies before you make your plans.

According to Tim Bouhour, director of product at Genomics England, successful product managers show ownership when they "proactively take things on before anyone comes asking. The kind of person who anticipates problems, demonstrates that they care about things and doesn't just wait for someone to ask or raise issues, is seen as far more trustworthy."

Another way to be proactive is to not only bring up problems when you find them, but also propose solutions. Of course, we all need help solving problems from time to time, but ownership means thinking through the problem and proposing a way that you can help solve it, rather than relying exclusively on others to come up with the answers.

Be wary, though, of being overly proactive with *other* people's problems. You may think you're being helpful, when really those people want the chance to come up with their own solutions. Focus first on the problems in your own orbit, and then offer, but don't insist, on helping other people with their problems.

Level 4: Mitigate risks

Nothing we do is without risk. So what are the risks inherent in your product or in your decisions? When we think about how customers will use our products, we build out possible scenarios or use cases. The process is similar when you make a decision or tackle a non-product problem. Here are some ways to identify risks:

- **Ask stakeholders and teammates:** Getting various perspectives on a situation can help you see different angles you may not have considered.
- **Learn from others' mistakes:** There may be something to glean from outcomes of similar decisions that have happened recently.
- **Brainstorm different scenarios:** Some risks are greater than others, but it's worth playing out different scenarios to see if there's something worth worrying about.

Once you have outlined the risks inherent in your plan, you can generate ideas for mitigating those risks. There may be no perfect solution, but having a plan in place shows ownership because you're not just making a decision and wiping your hands of it. Ownership is seeing the decision through to a successful outcome.

Level 5: Take responsibility

What remains consistent from the beginning is taking responsibility for your own mistakes and focusing on how to fix the problem and productively move forward. True ownership means not blaming others or getting defensive when mistakes are pointed out; rather, it is focusing on how to fix the problem, not who to blame.

If you are a people manager, you need to go beyond taking personal responsibility and take responsibility for your team. In *Extreme Ownership* (St. Martin's Press, 2017), Jocko Willink and Leif Babin describe leaders demonstrating ownership by taking responsibility for everything in their world, because there is no one else to blame. This fundamental mindset shift is a key way for leaders to build trust, both with their team members and also with stakeholders. This doesn't necessarily mean that you must fall on your own sword or resign when mistakes are made. But it does mean taking it upon yourself to make the necessary changes to prevent the mistakes from happening again.

Accepting feedback is ownership

We mentioned that taking ownership means not blaming others or getting defensive. Of course, as humans, our natural instinct is to reject any information that doesn't line up with the way we see the world, especially our role in it. This phenomenon has been coined "confirmation bias"* and to avoid it, we must seek out information from diverse sources, including those with whom we disagree.

In *Thanks for the Feedback* (Viking, 2014), Douglas Stone and Sheila Heen share that feedback is difficult for both the giver and the receiver. "But if the receiver isn't willing or able to absorb the feedback, then there's only so far persistence or even skillful delivery can go." Ownership is about understanding that none of us is perfect, and accepting that we all have opportunities for improvement.

To own our behavior, we have to really listen to feedback, not just hear it, because feedback is a gift. Stone and Heen advise that everyone should listen to feedback "even when it is off base, unfair, poorly delivered, and frankly, you're not in the mood." We might just find a nugget of insight that is true and helpful.

Melissa once had a manager who shared with her some hard-to-hear negative feedback. Melissa responded that the feedback couldn't possibly be true, because no one had ever told her about it before. Her manager replied with a question: "Is it that the feedback isn't true, or could it be simply that no one has ever told you about it before?" This completely changed Melissa's perspective, and made her realize that feedback can be a magical window into learning secret truths about yourself that everyone else already knows.

* Raymond S. Nickerson, "Confirmation Bias: A Ubiquitous Phenomenon in Many Guises," *Review of General Psychology*, 2 (2): June 1998, pp. 175–220.

Irie Takes Ownership

When Irie gets back from her walk with Sri, she brainstorms possible outcomes of Sparks promising all the AI product ideas to the board right away.

Irie looks at what she's written and decides that the biggest risk is if the board expects all of this quickly but does not provide additional resources to make it happen. They'd have to divert everyone to this AI effort, and that would be hugely disruptive to their current work. The design work would be shelved. There'd be no chance of doing the user access management features Arianna wants or any other large customer support work. *We have much more evidence that those things will add value than we do for any of this AI work*, she thinks. *So shifting 100% to AI is an enormous risk.*

She thinks this will be good enough to convince Divya of the need to set expectations properly with the board. She looks around the office, but Divya is not at her desk. She pings her on Slack and gets back a quick reply: "With Sparks."

Arriving at Sparks's office, she finds the door closed, but she can see through the window that Sparks is just finishing a meeting with Divya. They seem to be arguing as Divya stands up to leave. Through the closed door, Irie can just make out what they are saying.

Figure 4-5. Irie's map of possible outcomes

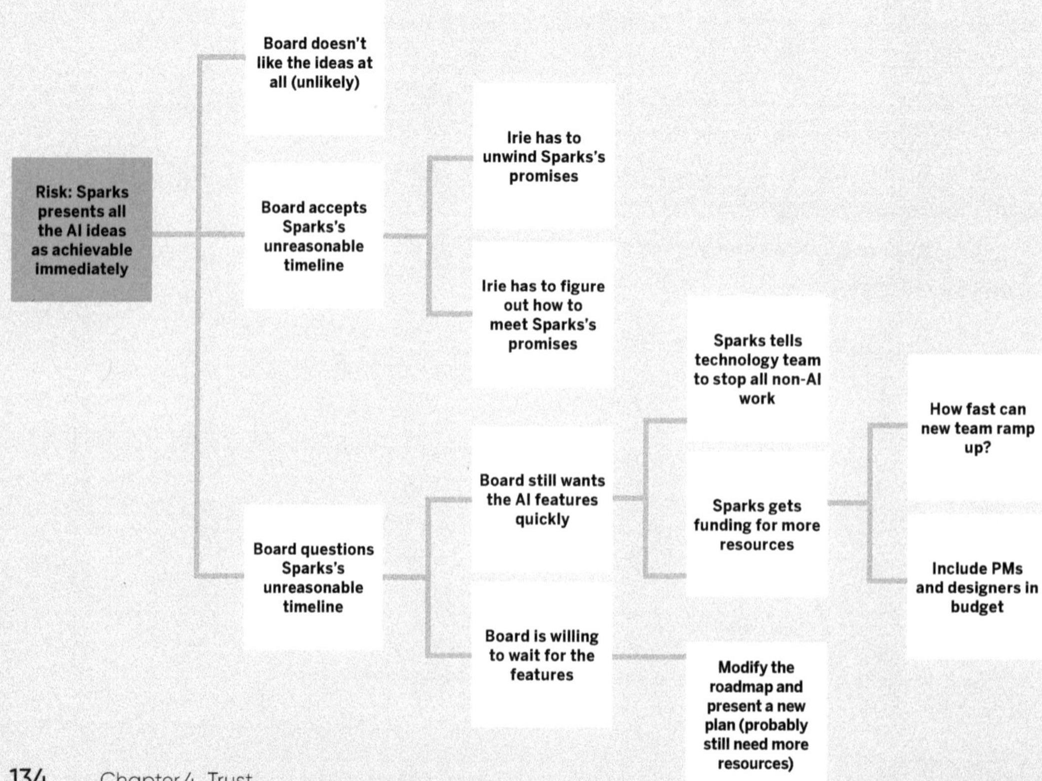

"I don't understand why you don't think we can handle it internally," says Divya. "We already have the expertise, we just need to expand the team."

"We can't move quickly enough," says Sparks. "This partnership would enable us to get started tomorrow."

"Sparks, it's not that simple," says Divya, seeming to balance her frustration with a desire not to get into an argument. "There's always ramp-up time to get someone onto a new project, partnership or otherwise."

"So wouldn't there be ramp-up time for a team in India too?" asks Sparks.

"Yes, but that's why I'm arguing they're equal. But if we build the team in India, we have more control over the intellectual property than if we partner with an outside company."

"I think we can agree to disagree on that one," says Sparks. "I have to hop on a call."

As Divya emerges from Sparks's office, Irie asks if she has a few minutes to discuss the board meeting tomorrow.

"Sure. But I haven't eaten all day. Mind if we chat over lunch?"

"That works. I haven't eaten either. I'll meet you in the kitchen," says Irie.

Fifteen minutes later, Divya and Irie are sitting with their lunches at a table in the kitchen. It's already past 1 p.m., so the kitchen is pretty empty. Irie has a sandwich she grabbed from the coffee shop downstairs. Divya brought leftovers from home.

"How was your weekend?" asks Irie, taking a bite of her sandwich.

"Not too bad. My daughter is finally healthy enough to go back to preschool today, so that's a relief. Feels like they send kids home for a single sneeze or cough these days."

"Tell me about it. Having kids these days makes it hard to have a job too," says Irie with a smile.

"Exactly," agrees Divya, giving in to a bit of a laugh. "Although I'd be happy to take some more time off, because Sparks is really driving me crazy."

"How so?" asks Irie, although she has an idea that it's related to the conversation she just overheard.

"He's working on some sort of partnership and he doesn't want my team to get involved," begins Divya. "It doesn't make sense. We can ramp up quickly with my connections in India, and labor there is still more affordable than in the US. We could easily build our own team in India, so I'm not sure why he wants to outsource. Can you talk some sense into him?"

"I'm flattered that you think I have so much sway with Sparks. I don't think he listens to anyone," says Irie.

"He seems to think highly of you. He told me about the ideas you sent him about possible AI work."

"But you and your team were part of that," Irie says. "You know all about it. Why is he telling you?"

Divya rolls her eyes. "Sparks doesn't get into the details of who does what, but he liked how you put the pieces together and made it easy to understand. I agree."

"Well, thanks, I appreciate that," says Irie. "But I was actually going to ask you for a favor. You're going to the board meeting tomorrow, right?"

"Yes. Sri can't make it, so he asked me to fill in."

"Well, maybe you can help me," says Irie. "You seem to have a good sense of how product development actually works—what's a reasonable timeframe on things, how feasible certain ideas are. Sparks thinks we can do everything immediately."

"So he's going to propose to the board to do all the AI ideas all at once, right away," Divya says. "Sounds exactly like Sparks. And that's a problem. If we dump everything else and focus on this, all my team's work for the last six months goes down the drain."

"Exactly," says Irie. "Maybe you can keep an eye on the conversation and make sure the board understands that we can't do everything at once."

"I can do that," says Divya. "But Sparks probably won't give me a chance to talk, and he gets mad when people try to interrupt him."

"Yeah, I've noticed that," says Irie.

"But wait, I have another idea," says Divya. "One of the board members is my former manager from my first job. He went on to be head of research at a Fortune 500 company and then CTO. He just joined the board recently. I've been meaning to reconnect. It might be a long shot, but I can try."

"That would be amazing!" says Irie. "And I can promise to have a conversation with Sparks about keeping the development in-house. I completely agree with keeping the IP in house. I can't promise the outcome, but I can try."

"Oh, you heard that," says Divya, reddening.

"I didn't mean to eavesdrop," says Irie, "I was looking for you and you were maybe a little bit loud."

"No problem," Divya says. "I should be more careful."

Irie wonders why advocating for insourcing would be sensitive, but decides to revisit that question another time. "I think if we work together on these things we'll be able to attack it from a few angles and get the best outcome for the company."

"I'll do my best," says Divya. ■

4.4 Irie Explains Dependability

The next morning, Irie is pacing the office, worried about the board meeting. She's not sure if she's done everything she can to fix the situation, but there's not much she can do now but wait. She starts reading the news on her phone while she paces. An article comes up about ethics in AI, and Irie is reminded of Sergey's concerns about job losses.

Irie writes Sergey a Slack message.

Hey Sergey, I thought about what you said about the negative impacts of AI, and I had an idea. Maybe we can put together an AI ethics committee before we get too far into the AI product work. That way we can put together a policy to guide what we do. We can identify risks and mitigations, etc.

She gets a response almost immediately.

That's a great idea. I think I heard Arianna express some concerns too. Maybe we can have a cross-functional committee.

Irie writes back.

Perfect. I'll see if she's interested and set something up.

As she's walking and typing on her phone, she almost bumps into Yacob.

"You seem anxious," says Yacob. "Everything okay?"

"Yeah, just juggling a lot of balls right now," says Irie.

"Actually, while I have you, I have one more ball to throw into the mix. I need to give you some feedback on Christina. I've been avoiding it, but I think literally bumping into you is a sign. Do you have a minute?"

"Sure," says Irie. "What's up?"

They move to Sri's office and Yacob closes the door. "I don't know if you've noticed in the few months you've been here, but Christina can be hard to work with."

"Tell me more," says Irie.

"Well, it's sometimes hard to get her to respond to emails or Slack. Sometimes a request will go unanswered for days."

"That doesn't sound good," says Irie.

"Mo has been complaining about it, but I don't think he's confronted her directly. Mo is not usually in the office because of the situation with his wife, so he can't just go find her at her desk. Even so, the relationship between a PM and an engineering lead should be pretty tight, so I'm not sure why she's not responding."

"Hm. Yeah, I heard something similar from Zola in support. I didn't realize it was that bad."

"I wouldn't have said anything, except I'm not sure how to fix the problem," says Yacob.

"I'm glad you did say something, and I appreciate your transparency," says Irie. "I'll talk to her."

Irie walks back to her desk and Slacks Mo. He's free, so she takes her laptop into a phone booth to call him. When he appears on screen, she relates what she heard from Yacob and asks for his input on Christina's behavior.

Mo looks uncomfortable. "I wasn't sure if I should say something, but it's hard to get her to respond to things. It's really slowing down the team. Plus, she'll only answer the exact question asked, so you have to ask a bunch of follow-up questions to get more information. It's kind of a time-suck."

"I'm sorry that this is happening. I'll talk to Christina and see what's going on. Have you talked to her directly about it?"

"No, there never seems to be an opportunity. I only ever see her in group meetings like standups and demos."

"Don't you have a one-on-one with her?" asks Irie.

"No, we never really set one up," says Mo.

"It sounds like you don't really have an opportunity to give her feedback, which becomes more difficult if she's not responsive to messages."

"Exactly," says Mo.

"Okay. I'll talk to her. Thanks for the context. But you really should set up a regular one-on-one. If you can't reach her to coordinate, you can just put it on her calendar and see if she accepts."

"I'll try," says Mo.

When Irie gets back to her desk, Arianna is waiting for her.

"Are you ever at your desk?" says Arianna, half angry, half joking.

"Not really," says Irie. She smiles and asks, "What can I do for you?"

"I wanted to see if you need anything for our call with RightBank tomorrow," says Arianna.

"No, I think I'm all set. We've already talked about their account history and what their concerns are. Mostly, I'm going to be asking them questions to get into more details about how they use the product and how that fits into the rest of their wellness programming."

"I can tell you all of that," says Arianna.

Irie explains that this is an opportunity for them to explain it all to her fresh in their own words. This will help them feel heard and validated. "I can send you the list of questions I plan to ask, if it will help you be more comfortable."

"Okay," says Arianna. "I'm still a little nervous. But I trust you. I believe you will come up with something brilliant to save the account!"

As Arianna walks away, Irie thinks to herself, *I appreciate the trust, but "something brilliant" is a lot to live up to. I hope I've set the right expectations.* ■

Dependability

At its core, dependability is simple: do what you said you were going to do. But in order to gain trust with your stakeholders, you need to think about a few more facets of dependability (Figure 4-6).

Figure 4-6. Facets of dependability

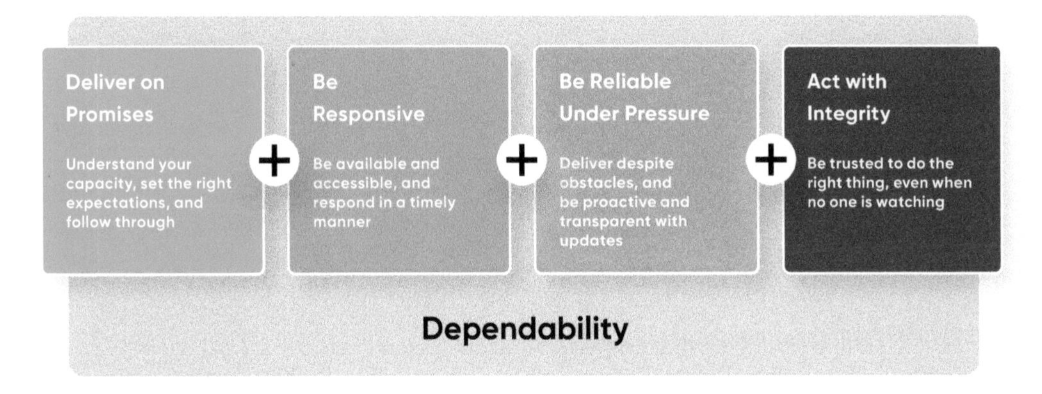

Deliver on promises

First and foremost, you have to deliver on your promises.* This involves setting the right expectations with stakeholders in the first place, and proactively updating them when things change.

Setting expectations and making estimates

To deliver on your promises, you have to make promises you believe you can achieve, thus setting the right expectations up front. This requires accurate estimates of your time and the time the team needs to complete their work. While engineers should be estimating engineering time, the product manager is the one who uses that information to create the roadmap, and is the face of the team to stakeholders. So the product manager needs to be confident in all estimates for the team.

* For products, "deliver" means to launch something into the market, like a product or a feature. In the personal sense, "deliver" means to follow through on whatever you personally promised, such as gathering information, creating a document, or executing on an agreed-upon decision. We will address both in this section.

Just like we add in buffers for anticipated usage levels, throughput capacity, and factors of safety, you can add in a buffer for engineering estimates, before reporting to stakeholders. Like the old adage: "underpromise and overdeliver." Not only does this make it easier to actually deliver on time (by expecting the unexpected), but if you have extra time (which does happen on occasion) you can throw in some "quick wins."

How much buffer you include depends on who is doing the estimating. Melissa once worked with a great engineering manager named Rusty. The problem was that Rusty was so good that he always thought the work would be easy for anyone. It might have been easy for him, but it wasn't necessarily easy for the rest of the team. So Melissa always added a 3x factor on his estimates, which ended up being fairly accurate. Figuring out your team's "Rusty Factor" is important for setting the right expectations with stakeholders.

Over time, you can build trust by getting better at estimation accuracy, which is critical when you have to tell a stakeholder that a particular solution will require more time to implement than they think it should. Keeping track of estimates and how you perform against estimates over time is one way to help refine estimate accuracy.

When you can't deliver

Sometimes you do miss a deadline, which erodes your dependability. The best way to handle this is to communicate it as soon as possible, even preemptively, like when missing a deadline is only a risk and not yet a reality. When you know that you will miss a deadline, use what you learned to create a new, more accurate deadline, and get alignment on the new plan. The credibility you gain from being transparent can help reduce a potential loss in assessments of your dependability.

As we'll see more in Chapter 6, sometimes you need to say no to things. The risk in saying yes to too many things is that you will not be able to deliver, which affects your dependability. That said, there are certain corporate cultures in which delivering 100% of what you promised is actually frowned upon—it means your goals were not ambitious enough. If you're lucky enough to work in one of these environments, you should get an idea of what the culturally approved delivery rate is—90%? 80%?—and aim for that. Any time you are aiming for less than 100%, it's useful to call out which goals you think you can hit and which are stretch goals. Or you can state confidence levels on each item. Either way, setting stakeholder expectations from the beginning is important.

Be responsive

The second key aspect of dependability is simply being responsive: being available for stakeholders and making sure they know how to reach you. This can be as simple as getting back to people promptly and following up when you said you would. This does not mean that you have to be available 24 hours a day, 7 days a week, and it does not mean that you have to give everyone your personal cell phone number. But it does mean that you need to respond within a reasonable amount of time.

One way to do this is to create your own personal Service-Level Agreement (SLA) either officially (shared with stakeholders), or unofficially (a general rule you set for yourself), so that stakeholders feel they can depend on you as a partner for whatever work you are doing together. Maybe your SLA is that you will respond to all messages within one business day. Maybe you will respond within one business day on Slack and within two to three business days on email. It's also helpful to be clear about what hours you're available, and who to go to if you're not around, especially in teams that span multiple time zones. Whatever you decide, you should be consistent and communicate your availability with stakeholders.

For example, if a stakeholder messages you with a question, respond as soon as you can, even if it's to say that you have a busy day and will get back to them tomorrow. And then actually find out the answer and get back to them tomorrow. These small interactions will build trust over time that your stakeholder can depend on you as their partner.

Be reliable under pressure

Obviously, it's easiest to deliver on your promises when everything goes as planned. But most of us do not live in that kind of world. Being dependable means that even if circumstances change, you can still meet expectations. You will be seen as dependable if you can handle unexpected challenges and adjustments.

There are unexpected changes, and then there are crisis situations like a product outage for customers, a security breach, or a major problem just before, during, or after a product release. As product managers, we have a unique position in the company as the nexus for cross-functional communication around the product, so we tend to lead in times of crisis. PMs speak the languages of the different stakeholders, so we are particularly helpful in an emergency situation, even if it's a highly technical problem that will likely be solved almost entirely by engineering.

The key is to remain calm and use your product management skills to logically assess the situation:

- **Gather information:** What information do you have about the crisis situation? What additional information do you need and how can you get it quickly?
- **Assess the problem:** What exactly is the problem that needs to be solved? How do you know? Think about the situation from multiple perspectives, e.g., engineering, marketing, sales, customers.
- **Identify the right people:** Who needs to be involved? Who is the decision maker? Who needs to be informed? What is the best way to communicate with them? Who is in charge of communication?
- **Define risks:** What are the risks and possible mitigations?
- **Design the solution:** What is the plan to get out of the crisis? What work needs to be done and by whom?
- **Monitor and retro:** How is the solution going? Did it fix the problem? How do you prevent it from happening again?

At a company Melissa worked for, when a product-related crisis emerged, the Product Team (product, engineering, and design) got in a "war room" to work out a solution. Product failures were critical at this supply-chain company because it meant that shipments could not be sent to customers. Critical outage communications were emailed to a specific internal mailing list and posted on Slack, with updates going out frequently—between every 15 minutes and once an hour. If it was a technical problem, most of the process was led by engineering, but anyone who wanted to volunteer could take on the tasks of updating stakeholders.

Act with integrity

Another form of pressure occurs when your integrity is tested. Acting with integrity means that you can be trusted to do the right thing, even when no one is watching. Start by defining your values, using your personal values, established corporate values, or both.

For example, Melissa worked at a company where one of the corporate values was winning together as a team. People who tried to climb the corporate ladder while kicking other people off the rungs were not acting with integrity according to the company, even though a team member may have personally valued being competitive more highly than being cooperative.

What about the questionable middle ground, where conflicts between values can arise? For example, what if you know something you aren't supposed to reveal but the other person would really benefit from knowing? Bruce was once asked directly by an employee if they would be laid off. He knew the answer was yes, but his boss told him explicitly that he could not talk about it. To make matters worse, he knew that this employee needed to make some time-sensitive financial decisions before he could officially share the news with them. He found a way to hint that they should be cautious by telling them that he couldn't answer that question at this time. He did this in a tone that made it clear there was information he knew, but couldn't tell them. Maybe he broke the company's (and his boss's) confidence. But he doesn't regret it.

Delivery Responsibility

Product delivery is a shared responsibility among the Product Team—product management, design, and engineering. This sharing requires you to build trust among these functions.

If responsibilities are unclear, you can specify them in a framework like RACI.* For example, you may agree with your teammates that engineering is responsible for creating and launching high-quality working code, design is responsible for the end product being usable by customers, and product management is responsible for the product solving a real problem in a valuable way.

Since the product manager is the key liaison with stakeholders, the lines between the responsibilities of different Product Team members can appear fuzzy. This is where ownership meets dependability. Because the PM is the representative for the entire Product Team, you will be making promises on behalf of the team, even if you are not personally doing the work (such as completing a piece of code on time).

Taking responsibility for a problem with delivery should be a joint effort. For example, if your team is about to miss a key deadline, taking ownership might look like you and the engineering lead presenting to stakeholders together on why the deadline was missed and how you, as a team, are going to fix the problem.

* RACI (Responsible, Accountable, Consulted, Informed) is a model that helps clarify the responsibilities of people or teams in collaborative endeavors.

Irie Exhibits Dependability

Irie and Christina are sitting together in a small conference room. Irie has a cup of coffee in front of her, and Christina has a can of diet ginger ale. Irie is asking Christina if she was able to get her car fixed.

"No, it's still in the shop. But at least I have a loaner car now. Makes things easier."

"That's good," says Irie. "Listen, I wanted to give you some feedback I've been hearing, and share some things I've observed. You let me know what you think, okay?"

"Uh, that doesn't sound good," says Christina.

"Every piece of feedback is an opportunity to grow, right?"

"I guess," says Christina, tentatively.

"I've heard from a few people that it's hard to get you to respond to requests, and I've seen it myself too," begins Irie. "Days might go by without an acknowledgment of a Slack message."

"Well, you know how busy we all are, right? With the volume of requests I get, I can't possibly keep up," says Christina, her voice rising. "I get messages at all hours of the day. Am I expected to be online and glued to Slack 24/7? That's unreasonable!"

"I'm not saying that," clarifies Irie, trying to convey calm with her voice. "I'm saying that you should at least tell people you've heard their request. You don't have to have an answer for it, but at least say something like 'Sure, I'll look into it.' Or 'I'm swamped right now, how urgent is this?' Something to let them know you've received their message."

"I guess that's fair," says Christina. "But even that can't be 24/7!"

"No, of course not," says Irie. "Can you think of a way that you can consolidate the work of doing the initial responses?"

"Maybe..." begins Christina, thinking about it. "Maybe first thing in the morning each day I can go through any requests from overnight and just let them know I got it."

"That's a good start," says Irie.

"But even if I do that, how am I supposed to actually follow up on all that stuff?" asks Christina. "I get a lot of requests. Some of them are from everyday stakeholders, like José or Divya. And now that I think about it, I think I do owe Divya a response to her question about estimates for implementing the action propensity model." She makes a note in her phone to follow up with Divya. Then she returns to the conversation. "But most of the requests are from sales for one-off customer requests! Am I supposed to say yes to everyone?"

"You definitely should not be saying yes to everyone," says Irie. "But too many requests from sales is not a problem with you, it's a problem with all of us, and I'm working on fixing it. Maybe we can find a solution together for a triage process, to unburden you and Eitan and Min from having to deal with individual sales requests all the time."

"That would be amazing," says Christina. "And it would be a lot easier for me to manage my messages, if we remove sales from the equation."

"Okay, let's work on this as a group in our team meeting next week. But for responding to people, does what I said make sense? You don't have to be online 24/7, but at least acknowledge receipt of messages by, say, the following business day?"

Christina agrees and Irie takes a moment to try to summarize. "When you take ownership of a request, it shows that you are dependable, and people can trust you to deliver on your promises. If they don't trust you, they come to me. Being dependable helps you build better stakeholder relationships, and, selfishly, it makes my job easier."

Christina thinks for a minute, then makes a proposal. "Let's say we take the sales requests and funnel them through some sort of triage process. Then every morning, I look through any outstanding messages and acknowledge them. And then I can triage and follow up on things based on urgency and availability. Does that sound right?"

"I love how logically you think through things," says Irie. "That sounds right to me. But the follow-up still seems a bit open-ended. How about we add two more things: 1) No one should have to ask you for an update more than once before you respond, and 2) No one should have to escalate to me because you haven't responded yet. That's how we'll know you're succeeding with this plan."

"Okay, I think I can do this," says Christina.

The next day, Irie dials into the RightBank meeting. She is working from home, which is just as well because the three people from the customer side are also dialing in from home. Irie is a bit preoccupied because she still hasn't heard anything about the board meeting. But she tries to put that out of her head as she joins the meeting.

Arianna begins. "Hi everyone, thank you for joining me. We have a very special guest today, Irie from the product team. Irie, please introduce yourself."

"Hi all. I'm Irie. I run the product management team here at Helthex. My job is to make sure that our product solves your problems, which is why I'm here—to make sure I understand your needs."

"Thanks, Irie, that's helpful," says Joe, a program manager for the benefits program at RightBank.

Irie continues, "Arianna told me that she already shared the news about user access management not being ready for a while, and I'm sorry we can't get it for you right away."

"It's disappointing, but I guess there's nothing I can do about it," says Joe, crossing his arms in front of him. "But Arianna said you have some other ideas and some new features to show us."

"Yes," says Irie, "we do have a number of new features we are releasing soon. But first, I'd like to get a better understanding of how you use our current product with your wellness benefits program."

Irie asks the folks on the call a number of questions about their wellness programs and the problems they're facing. She discovers that their main concern right now is the arduous manual entry task of adding new accounts and deactivating old ones. They have only two people who do this, so they don't need a complicated permissioning hierarchy or a dedicated UI, at least not right way. She shares some mockups of the bulk upload functionality the team is working on, and Joe is impressed. Irie assures him that more robust user access management is something they are working on down the road.

"I didn't realize you were already working on it," says Joe. "It's not the full solution we asked for, but it's good enough for now, and I appreciate the time you took to understand my frustrations. Arianna, thanks for bringing Irie on the call today, I feel a lot more comfortable about waiting on full user access management. And I'd like to get the bulk user upload feature as soon as it's ready."

"We're looking for some more beta testers, if you're interested," says Irie. "Then you could tell us what you think before we release it for everyone."

"That would be perfect," says Joe.

After they end the call, Arianna texts Irie:

Thank you! That was a great call. You really came through for them. Joe already okayed the renewal!

Irie replies:

Just doing my job.

Irie's satisfaction is short-lived, however. A message from Sri appears on her phone:

Ping me as soon as you can.

This is immediately followed by:

Sparks is on the warpath. ■

Takeaways

Cultivating your stakeholders' trust in your work and abilities is critical to your personal success, and the success of your product. There are some key ways you can demonstrate your trustworthiness:

- The best way to become a trusted expert is to "show, don't tell." Speak your stakeholder's language to demonstrate an understanding of their world, share relevant experience to help address problems, and provide targeted education by showing how you can help meet your stakeholder's needs.
- If you have confidence in yourself, believing in your own abilities, your stakeholders will have more confidence in you. Confidence doesn't mean being pushy or assertive; you can build it proactively, by being prepared when you speak with your stakeholders and providing transparency into your processes.
- Own your responsibilities to gain trust. There are five levels of ownership: execute well, involve stakeholders, be proactive, mitigate risks, and take responsibility. Ownership also depends on your ability to take feedback constructively and without becoming defensive.
- To be trusted as a dependable partner, be diligent about delivering on promises, be responsive in communication, be reliable under pressure, and always act with integrity.

A roadmap that people can rally around requires buy-in on objectives

Chapter 5

Roadmap

If you've successfully implemented the recommendations in the first four chapters of this book, you've discovered the shape and culture of your organization, you've identified your key stakeholders, and you've established rapport and trust with them.

Even after building strong connections with your stakeholders, you've likely found that they still demand more than your team can deliver in any reasonable timeframe, and you need a way to sort through conflicting inputs and priorities. In this chapter, we will discuss how to bring stakeholders together to craft a set of shared objectives and a product roadmap.

As Irie tries to drive alignment, we will discuss how to:

- Derive product objectives from organizational objectives.
- Use workshops to drive alignment.
- Mine for conflict, uncovering hidden misalignments.
- Develop a product roadmap that focuses on customer needs and product objectives.

We begin as Irie confirms her suspicion that there is no consensus on objectives among her executive team and goes about trying to establish a firmer foundation she can build her product roadmap on.

5.1 Irie Seeks Organizational Objectives

"You handled the situation with Arianna well," Sri says to Irie after she brings him up to date.

"She seems happy," Irie agrees, "but we've just postponed the problem. We'll need to do user access management eventually, but we don't have enough engineering resources to do that and the AI work at the same time. I'm guessing your note about Sparks being on the warpath means we're committing to that?"

"I don't think it's quite that bad," begins Sri. "Sparks kicked Divya out of the board meeting after the staffing discussion, so she didn't hear the AI conversation. She managed to connect with her friend on the board, though, and I gather he asked a lot of our questions for her."

"Well, that sounds great!" Irie says. "Is that what Sparks is upset about?"

"Yes," Sri replies. "Divya's friend knew enough about it that Sparks figured out that he'd been talking with someone here. He wouldn't reveal his source, but Sparks is feeling like we undermined him with the board."

"And he wants someone to blame," Irie concludes.

"He asked me directly who had been speaking to the board," explains Sri. "I had to tell him it was Divya, but I told him she did it because I asked her to."

"Sri," Irie says, "it was much closer to me asking her. I mean it was her idea, but she was trying to help me."

"She was helping all of us," Sri says firmly. "And so was her friend on the board. I pointed that out to Sparks and I hope he heard me." Sri smiles ruefully. "And then I went home and watched a cricket match. I hate confrontations like that."

"So what do we do now?" Irie asks.

"We're at a stalemate," Sri says. "We have design work in progress, sales wants one thing, Sparks wants something else. I wish Liz were here," he adds. "She'd know how to resolve this."

"Is she completely out of the loop? Irie asks. "Can't we ask her to weigh in on strategy?"

"I asked Sparks that, too. He says we need to respect that she's on leave. I have to confess," Sri adds, "I find myself wondering if he's using her absence as an excuse to dictate priorities."

"That crossed my mind," Irie says. "But maybe there is a way we can figure out 'what Liz would do,' to borrow Christina's favorite phrase. It feels like the real reason we don't have a roadmap is what we discussed the other day: we don't have clear objectives. We don't even know if we're supposed to be profitable. Did you know that our corporate deals are not profitable, on average?" Irie asks.

"We're a growth-stage company," Sri replies. "I didn't think profit mattered until we hit scale."

"Finance seems to think it matters," replies Irie. "They hired Justin to analyze the profitability of every deal. And it gets worse," she continues. "I'm working on how to triage requests from around the organization. The data is messy, but it looks like half of what we've done over the last year is based on corporate customer requests or escalations."

"Wow," says Sri. "So we're spending half of our time working on stuff for unprofitable customers? No wonder we aren't making progress on the roadmap! I'm going to speak to Ella about this nonsense with sales dictating our priorities," Sri continues.

"Sales is just doing what they are paid to do," Irie says. "It's up to us to sort out the priorities for the product based on goals. The trouble is finance and sales don't seem to agree on what the goals are. Don't we have some sort of organizational objectives?"

"We have a budget," he offers. "And we created a bunch of OKRs about six months ago, shortly before you started here. But to be honest, we haven't discussed them since then."

Sri opens his laptop, searches for a moment, and then shows Irie a slide showing the company's objectives and key results (OKRs).*

Figure 5-1. Helthex OKRs

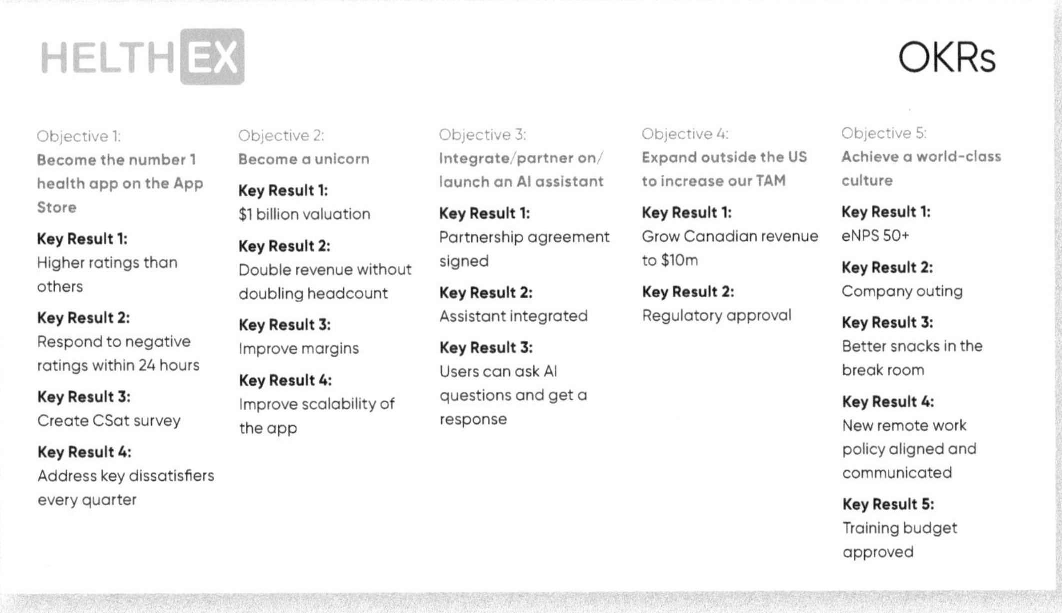

HELTHEX

OKRs

Objective 1:
Become the number 1 health app on the App Store

Key Result 1: Higher ratings than others

Key Result 2: Respond to negative ratings within 24 hours

Key Result 3: Create CSat survey

Key Result 4: Address key dissatisfiers every quarter

Objective 2:
Become a unicorn

Key Result 1: $1 billion valuation

Key Result 2: Double revenue without doubling headcount

Key Result 3: Improve margins

Key Result 4: Improve scalability of the app

Objective 3:
Integrate/partner on/launch an AI assistant

Key Result 1: Partnership agreement signed

Key Result 2: Assistant integrated

Key Result 3: Users can ask AI questions and get a response

Objective 4:
Expand outside the US to increase our TAM

Key Result 1: Grow Canadian revenue to $10m

Key Result 2: Regulatory approval

Objective 5:
Achieve a world-class culture

Key Result 1: eNPS 50+

Key Result 2: Company outing

Key Result 3: Better snacks in the break room

Key Result 4: New remote work policy aligned and communicated

Key Result 5: Training budget approved

---

* Authors' note: These are truly terrible objectives. Do not write objectives like these. For guidance on how to make great objectives using the popular OKR (Objectives and Key Results) framework, see Christina Wodtke's excellent book *Radical Focus* (Cucina Media, 2015).

Irie reads the slide with some difficulty due to the small text. "Wow, these are a lot of objectives. I see growth and revenue and margin, some product initiatives...and a lot of other things. Were we really supposed to be able to achieve all of this in a year?"

Sri says "This is only slide one, here's the second slide."

Irie falls silent, dumbfounded.

"Some consultants told us to be 'aspirational,'" Sri says, sounding a little defensive, "so we wrote down everything we wanted to shoot for."

Irie sees the objective about hiring a VP of product and asks Sri about that. "That's you," he explains. "We couldn't afford the salaries we were hearing about so we decided to go for a director."

Unsure what to do with this information, Irie refocuses on how to apply these OKRs to her problem. "I'm not sure where to start in creating objectives for the product," she says.

"I see a lot of things for the technology team here," says Sri, pointing to App Store ratings, scalability, the AI assistant, and new features like work for Nike.

Irie sits back in her chair, trying to decide how to extract something useful for her team from these OKRs. It's one of the worst sets of objectives she's ever seen. She doesn't want to appear too critical, though, so she speaks tactfully. "I agree that some of these things affect my team and the product," she begins. "But I don't normally like having specific features as objectives or key results, not unless they are a legal or regulatory requirement. Normally we would focus on the customer

Figure 5-2. Helthex OKRs, continued

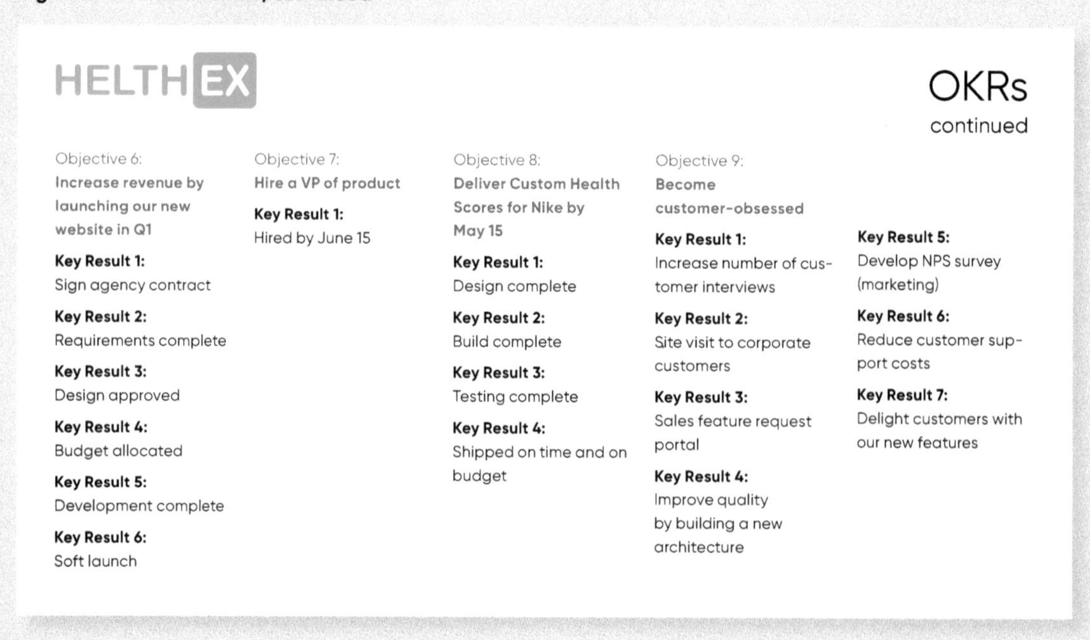

behavior and other things we can test and measure. We can get to that," she adds hastily, "but first I need to understand which of these organizational objectives are most important."

Sri says that all of these organizational objectives seem important, so Irie proposes an analogy. "If I'm traveling with my family, and I have to pick a place for dinner, what criteria do I use to choose a restaurant? Aside from the type of cuisine, I could consider price or distance or star ratings."

Sri nods, so Irie continues. "I'd love it if the best-rated restaurant in town were also the least expensive and located next to our hotel, of course. But since that's pretty unlikely, I need to know which is the most important objective to know when it's okay to ignore the budget in favor of something else and when it's not."

"So you would need to know if the revenue goals are more important than the profitability goals to make a decision on product priorities?" asks Sri. "Like whether to do user access management first or AI?"

"Right. We can't align our product strategy with nine different objectives and all these KRs," she adds. "If I'm going to come up with a product strategy and a roadmap that people can rally around, I need buy-in on which of these objectives we should really be focused on. Then we can create some more specific product objectives that align with those. Then," she adds, "we'd figure out which features or other things are most likely to help us achieve them."

"I like this idea of traceability from organizational objectives to product objectives and then to features," Sri says. "It's elegant and it appeals to the engineer in me. But how are we going to get everyone to agree on where to focus?" Sri asks. "We tried six months ago with this list and we came out of it with nine number one objectives."

"I think we've established good enough relationships with our key stakeholders to have a workshop where we talk about this stuff openly."

"We can talk," Sri says sardonically. "That doesn't mean we'll get anywhere."

"When Darius needed alignment with the executives on something, he'd get all the players in a room to hash it out," Irie explains. "I did the same thing with my stakeholders, but Darius was a pro. He'd ask a lot of questions and get people to come to their own conclusions before driving toward a decision. My friends and I recorded his bag of workshopping tricks in our playbook." ■

Workshopping Drives Alignment

Workshopping is a technique for collaborating with a group of people to create something together. It's used by the most effective product leaders to generate ideas and solicit feedback. Most importantly, workshopping is a particularly good way to gain alignment.

Involving your stakeholders in the decision-making process creates a kind of buy-in that is not possible when one person simply takes charge and announces a plan.* The old trick of getting people to accept a plan by making them think it was their idea (or actually making it their idea) is not only good advice, it's rooted in science.

In their book, *The Human Element* (Wiley 2021), Loran Nordgren and David Schonthal describe an experiment conducted in the Harwood manufacturing plant by a man named Alfred Marrow. In the experiment, workers were split into three groups. Each group was asked to adopt a new process, but under different conditions: 1) training, where the workers were simply given instructions, 2) training with feedback, where workers were given instructions and were asked for feedback and adjustments were made before adopting the new process, and 3) codesign, where the workers were given the problem to solve and participated in the creation of the new process.

You may have guessed that the codesign condition was the most successful. Using that approach, the workers were more productive than before, and worker/manager relationships were better. Perhaps surprisingly, though, the other two conditions, training and training with feedback, were almost indistinguishable from each other. Both drove lower productivity than before the process change, vocal resistance to the change, and poorer worker/manager relationships.

We can learn from this that simply communicating your objectives, priorities, or roadmap, without involving your stakeholders in the creation of the plan, drives resistance to the plan. And simply asking for feedback on your plan does not alleviate this resistance. You need to bring your stakeholders with you on the journey, and make them feel partially responsible for reaching the destination successfully.

---

* We discuss Participative decision making in Chapter 1, Section 3: "Irie Learns About Decision Making."

Workshopping involves gathering people (in person or virtually) and going through a series of activities together that lead to a collective decision or other output. A successful workshop requires a few key elements, including a clear purpose, the right participants, ground rules, and productive activities.

Workshop purpose

The purpose of your workshop could be *generative*, like listing problems or coming up with solution ideas, or it could be *evaluative*, like taking everything you've learned in a series of customer interviews and looking for patterns. The purpose could also be *decisive*, like setting a direction, a goal, or a set of priorities. If needed, a workshop can target all three.

Co-create your workshop by asking participants to help define the purpose, activities, and ground rules.

Before you invite people to a workshop, explain your goals, the expected output, and what you need from them. If they understand and share your purpose, they'll be more enthusiastic participants.

Workshop participants

When choosing who to include in your workshop, ask yourself these questions: Who has relevant and useful information? Who might object later if they aren't included? Which Power Players might you want to include?* Which functions are affected?

Workshops often involve a trade-off between including all the relevant people and limiting the list to drive focus. The ideal number of participants for a workshop depends on its purpose. For generative or evaluative workshops, larger groups from ten to twenty people can create a lot of ideas and engaging discussion without becoming unmanageable. For decisive workshops, however, smaller groups of three to five people are better, because smaller numbers increase your chances of making a firm decision.†

---

* See Chapter 1, Section 2: "Irie Identifies Power Players."

† You can read about the importance of group size in "7 Strategies for Better Group Decision-Making," by Torben Emmerling and Duncan Rooders, *Harvard Business Review*, September 2020. *https://hbr.org/2020/09/7-strategies-for-better-group-decision-making*.

If your workshop has multiple purposes, consider holding several sessions, each with its own goal and participants. You could then, for example, ask a large, diverse group to generate ideas and summarize them, then ask a representative subset to choose and refine the best among those ideas.*

Sometimes your stakeholder either doesn't have time to attend the workshop, or isn't convinced it's a good use of their time. In these cases, it's helpful to meet with the stakeholder before the workshop, to get their input, and after the workshop, to give them an overview of the outcomes. While that may not feel like a good use of your time, these one-on-one sessions can be extremely valuable to get hesitant stakeholders on board with the process. If you can't find any time with them, you can email or message them. You can also ask them to send a delegate in their place.

Workshop ground rules

A workshop can set a group of people on a great path forward, but it can also veer off course if there are too many tangential conversations, attendees who don't participate, or disrespectful behavior.

You can encourage frank but respectful engagement by setting ground rules. We have found the following ground rules to be useful. It can be helpful to begin your workshop by going over the ground rules. In the spirit of co-creation, you can even ask if any attendees would like to add to the list.

* We learned a lot from *Design Sprint* by Richard Banfield, C. Todd Lombardo, and Trace Wax (O'Reilly, 2015).

Figure 5-3. Sample workshop ground rules

Disagree
Share what you think honestly, even if you believe others will disagree.

Assume positive intent*
We all want to solve this problem, even if we approach it differently.

Be curious
If someone says something you disagree with, seek honestly to understand their reasons.

Strong opinions, loosely held
Be open to changing your mind.

Be respectful
Focus on the merits of ideas, not the merits of people.

Make "I" statements, not "you" statements
"Here's how I see it," rather than "you're wrong."

Start with agreement
Instead of "no," say "yes, and" to build on others' ideas.

Admit your biases
If you prefer an option for personal reasons, it's okay to say so.

Raise your hand and don't interrupt
Everyone is expected to speak. Everyone is expected to listen.

No multitasking
Everyone is expected to be mentally present for the work.

One conversation
Everyone should hear what you've got to say.

Commit
Leave disagreements behind after decisions are made in the workshop

* The coaching team at Neuberg Gore offer some great tips in "A CEO's Advice: Assume Positive Intent," Neuberg Gore blog, accessed on February 5, 2023, *https://oreil.ly/Ls5kd*.

Workshop activities

Successful workshops typically consist of a series of structured activities, each with an intended output. These activities often build on one another, such as going step by step through the roadmap creation process, or discussing pros and cons for new potential markets. For a tough business problem, for example, you might have an exercise to define a problem, followed by a brainstorm to develop solutions, and finally a decision-making process to select one of those solutions to test.Here are some of our favorite workshop activities and when to use them (Figure 5-4).

Figure 5-4. Useful workshop activities

Silent writing

Each participant takes a few minutes to write down their ideas before discussion. Each shares what they've written out loud.

When to use

When you want to source multiple ideas or opinions, this technique gathers everyone's input before they are influenced by others. Particularly useful when a few voices tend to dominate or some people are reluctant to speak up.

Pro tips

Some teams prefer to do silent writing as pre-work, others in the first few minutes of the exercise. Figure out which approach works for your team and be clear about the expectation.

Breakouts

Participants break into small groups of three to five to work on a problem together. Each group then reports back to the full workshop group on their results.

When to use

When you want to ensure everyone in a large group has a chance to contribute. Also useful to parallelize discussions of multiple problems. Especially useful when there are natural groupings by expertise or existing teams.

Pro tips

Maximize diversity within each breakout group by including a variety of functions, seniority levels, genders, etc. Keep groups between three and five people for discussion.

Group critique

Participants provide feedback on the ideas presented. They are encouraged to begin with what they like about each idea before offering suggestions for improving them.

When to use

When you want to improve a set of ideas or proposals. Particularly useful to refine ideas between breakout sessions. Also useful prior to dot voting to be sure all ideas are understood and refined.

Pro tips

To avoid biases, don't put people's names on the ideas. You can use a framework of "I like" and "I wish" for the pros and cons of each idea.

Affinity mapping

Participants post their ideas anonymously in a common space, e.g., stickies on a wall or Miro board. Participants then silently arrange the stickies into groups of similar ideas, then name the theme for each group.

When to use

When you have a complex set of issues or a large and diverse group of opinions. Particularly helpful in identifying patterns and commonalities in a large group.

Pro tips

Find ways to make it less obvious which ideas came from which person, e.g., use one color of sticky for everyone and the same color pen or marker. Try to keep the group size to six to eight people max.

Dot voting

Participants are given a fixed number of votes they can apply anonymously to a set of competing ideas. These "dots"* can be placed all on one idea or spread out. Tally the votes to determine the highest priority ideas.

When to use

When you have a large set of ideas and you want to identify the most promising or important. Particularly useful as a follow-up to silent writing and affinity mapping.

Pro tips

Don't give too many votes to people. Force them to prioritize by giving fewer votes than options, e.g., three votes per person to apply to a list of five to seven options.

1, 2, 4, all

Participants write down their ideas on their own first, then discuss their ideas in pairs, then in groups of four. Finally, each group of four appoints a spokesperson to share with everyone.

When to use

When you want to iteratively gather a variety of ideas and drive toward alignment, while generating a cooperative atmosphere among participants.

Pro tips

Make each round quick so that discussion doesn't drag and get repetitive.

* The "dots" are traditionally applied with marker to sticky notes in in-person workshops. Miro and other online collaboration tools have similar functionality for remote workshops.

Additional workshop facilitation tips

Here is some additional advice we've found useful based on years of facilitating workshops.*

- Choose a single facilitator for the workshop, which makes it easier to understand who is responsible for moving the conversation forward. You can also choose a helper who can keep time, take notes, and assist with the activities.
- Review the goals and agenda at the start of the workshop and at key points throughout the session. Discuss adjustments to the scope and time allocation openly with the participants, adjusting course as needed. This usually increases engagement and the likelihood of success.
- Set up a "parking lot" list of topics that should be saved for another day. This helps keep the discussion on track. Make the parking lot list visible to all participants, and announce when a topic is being added to it.
- Encourage input from everyone by using silent writing and dot voting (Figure 5-4) or by directly calling on quieter participants. This ensures diversity of inputs and prevents silent dissent from undermining alignment.
- Have a code word like "ELMO" (Enough said, Let's Move On) anyone can use to refocus discussion when it strays off the main agenda. You can add tangential topics to the parking lot.
- Have a code word or a hand gesture (like a "time out" sign) when someone feels the discussion is getting unnecessarily combative.
- Use a large clock to time sessions so that everyone can see it.†
- Combine online collaboration tools like Miro or Mural with video conferencing tools like Zoom, Teams, or Google Meet to conduct virtual workshops.‡
- Some teams prefer to communicate in long-form memos, some in bullets, and others in visuals like diagrams or flow charts. Ask your stakeholders directly what works for them and try to use their preferred medium as a default.

---

* The new book from Rob Fitzpatrick and Devin Hunt, *The Workshop Survival Guide* (Robfitz Limited, 2019) is a great resource.

† Our favorite is called a "Time Timer," *https://www.timetimer.com*.

‡ Many people have learned how to use these tools in the last few years, but not all. Consider a quick refresher exercise with the tools as pre-work or a fun ice breaker.

Irie Workshops Objectives

Irie and Sri agree they need a workshop to align the key players around common objectives that will inform product strategy. Irie looks over her Stakeholder Canvas and proposes they ask Sparks, Philippe, and Ella to join them. Each is a Power Player, and among them they represent all key functions of the company.

Irie wants Sri to lead the discussion since he is the executive for the technology team, but he insists she lead. "You have more experience than I do in leading workshops, and you can be a mediator if we start to argue," he says. "I'll be the stakeholder for technology, but you should lead this." Knowing how Sri feels about running meetings, she agrees. She isn't sure if this is an opportunity to shine or to go down in flames, but she decides to take on the challenge.

Unexpectedly, the first challenge surfaces before the workshop even begins. Irie is in Sparks's office discussing her team's progress on vetting the list of AI ideas. "We think if we test these ideas with potential users before development begins, we can home in on the most promising ones and focus on those," Irie explains. "This will speed up the timeline and reduce risk."

"That's fine," Sparks says without seeming to listen. "I want you to use Alvex's AI tech for this. That'll speed things up even more."

"It might get us to market faster with something," Irie begins, "but Divya—"

"I know what Divya thinks," says Sparks, cutting her off sharply. "She wants to build an empire for her friends and family in India, but we have much more pressing concerns here."

Divya's motives aside, Irie feels it makes sense for the company to own the technology. Following up on Sparks's use of the word "pressing," she asks, "Is this just about speed?"

"What do you mean?" Sparks asks, looking directly at Irie for the first time in the conversation.

"It seems like there are a lot of factors to consider," she says. "Faster to market is good, of course, but focusing on the use cases that will drive adoption is important, too. And so is having an underlying technology that performs and scales. And that's after we decide if we're prioritizing AI, usability, or enabling corporate customers with things like user access management."

Sparks's nostrils flare as he speaks forcefully. "I know you and Divya and Sri want to doodle on your whiteboards about a perfect solution in a perfect world until we're all old and gray, but I have a company to run. I don't have time to debate every detail. I am in charge while Liz is out. I am telling you that the AI work is top priority. And I am telling you to use Alvex to make it happen ASAP. Is that clear?"

Sparks's diatribe reminds Irie forcefully of his Dominant decision-making style. Overwhelming him with all of the facets of the decision was a mistake, she thinks. She decides to change her approach.

"I'm sorry, yes, that's clear," she says. "Without slowing things down, though," she adds, "it might still be good to put together a proper roadmap with objectives and an idea of timelines."

"Is that what this meeting Sri called is about?" Sparks says, waving a hand dismissively.

"Yes," Irie confirms.

Looking at the calendar on his laptop, Sparks asks, "Why do we need Ella and Philippe in this meeting?"

"We want to come out of this with everyone aligned behind the plan," Irie says, "so we can execute efficiently and effectively as one team."

Sparks seems to be listening, so Irie continues. "I agree too much analysis can slow things down, but if we get their input on things we can put any concerns to bed now. And even the best ideas should be pressure tested, right?"

"All right," says Sparks. "If you can get people in line, I'll go along. But I need you to keep this process from spinning into an endless debate."

Once again, Irie wonders what she's gotten herself into.

A week later, Irie and Sri are in a conference room while Ella, Philippe, and Sparks are on video. Irie and Sri have joined the meeting on their laptops as well, making everyone into a talking head on everyone else's screen, leveling the playing field. They spend a few minutes fiddling with their sound controls to avoid feedback before deciding to sit side-by-side and mute Sri's mic.

Sri opens the meeting. "As part of her job as director of product, I've asked Irie to develop our product strategy, and she needs your help with the objectives."

"The objective is growth," Sparks says, in a tone suggesting that this is obvious.

Irie is about to follow up when Philippe jumps in, "We also need to be profitable."

Sparks is about to respond when Sri adds, "And I would like to make sure we take enough time to build a quality product."

Ella chimes in, saying, "My team will go wherever they're sent, but we'd sure like to win a few."

The room goes silent for a moment as the realization sinks in that they each have their own objectives.

"This is why we wanted to talk," Irie explains. "I read the company OKRs. You've all just mentioned a few of them. Before we can create a product strategy, we need to align on our objectives. What do we want? I cannot develop a product strategy for a set of incompatible goals."

"Who says they're incompatible?" asks Sparks, sounding indignant.

"Growth and profitability are always in tension," chimes in Philippe. "They hold each other back, like they are tied together by a rope."

"Particularly when we have to discount to win," Ella adds, "and when our corporate renewals suck."

"Well, that's technology's responsibility," says Sparks. "If they would ship what customers need, we'd win more deals, right?" He looks directly at Sri and Irie. "I'm sorry, but that's what we're really talking about here. If you two could keep up, we could win more deals and we'd have growth AND profit."

Irie knows data isn't Sparks's style, but she wants to show Ella, Sri, and particularly Philippe that she's done her homework. She shares a slide showing that her team has increased their output significantly in the last six months. "55% of our feature work has gone into requests from corporate customers or prospects," she explains. "And nearly 75% of our resolved bugs are from these customers."

"So it's a quality problem." Sparks states flatly.

"No," replies Sri sharply. "My team is solid, but they have to rush through things to respond to deals in the pipeline."

"What if I could get you more developers?" asks Sparks, leaning into his camera.

Sri responds that this would, of course, help, but that he is still concerned about lack of focus. "If you want us to go faster, we have to first go slower to get things right before we commit them to customers."

Irie decides to get the conversation back on track. "These are all great points, and it feels like everybody here is right," she begins. "We do want growth and profit, a great win rate and renewals, more great features, and quality. But whether we hire more people or not," she continues, "we'll always have finite resources, so we have to allocate those resources wisely."

Irie explains that she'd like to use an online collaborative whiteboard tool for their work together. She invites them to the shared workspace and, once they've all logged in, she demonstrates what she has in mind by creating three virtual sticky notes, placing them side by side for them all to see. They are: profit, growth, and quality.

Profit	**Growth**	**Quality**

"Based on what I've heard, we have these competing goals," Irie says.

"What about new features?" asks Sparks.

"New features aren't a goal in themselves, surely," says Sri.

"Agreed," says Philippe. "I'm interested in features if they drive growth or profit."

"It's important to make a clear distinction between outputs, like features, and outcomes, the results those features create," Irie explains.

Sri chimes in quickly, "So bug fixing or performance improvements would be outputs that could result in an outcome of improved quality."

Irie agrees. "Yes, or even design enhancements could be perceived by users as improved quality."

"What about win rate," asks Ella, "or retention? Those seem like outcomes, not outputs."

Philippe says that he thinks those would fit in underneath growth and profitability. "If we have a better win rate and better retention, we'll grow faster. And the cost of each sale will decline as a percentage of revenue, which will boost margin."

"Okay, so to keep it simple, can we say that 'profit, growth, and quality' are the three goals we care about at the top level?" Irie asks.

"Hang on," says Sparks. "These are all fine, but I still say growth is the most important. At the end of the day, that's what the market cares about."

"Maybe so," says Philippe, "but it's no good if we run out of money."

Ella leans heavily back into her chair, looks at the ceiling, and says, "Here we go again."

Irie looks to Sri, who is looking back at her with what seems like expectation. "I have an idea," says Irie. She fiddles with some controls in the virtual white board and then explains, "Everyone gets three votes. I'll set a timer and you can place dots on these goals for your votes. Spread them around between these three or bunch them up however you like."

Ella, Philippe, and Sri readily comply, placing their dots under the words on the virtual stickies. Ella moves quickly, adding two dots to growth and one to profit. Philippe places two dots under profit and one under growth. Sri places all three under quality. "I think quality drives the other two in the long run," he says.

Sparks waits until the others have gone, then weighs in with a sigh, clearly humoring them all. He places three dots under growth and sits back from his camera.

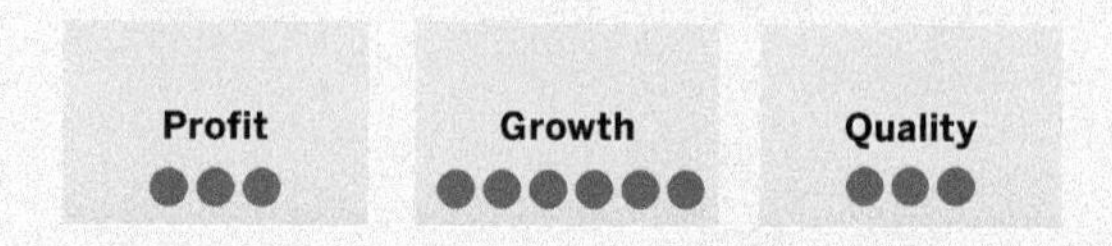

"What about Irie's vote?" asks Sri.

"Why does she get a vote?" Sparks demands. "If we're going to bring in non-execs, then I want Sergey and Arianna to vote."

"It's all right," says Irie. "My votes wouldn't change this picture. I agree with it. We've prioritized growth, followed by profit and quality."

"See, I told you growth is what counts," says Sparks smugly.

"They all count," says Ella. "Remember all those OKRs we did at the annual kickoff? These are the top three." She turns to Irie. "So what do we do with these priorities? We still have three goals."

Irie explains that she's used various methods of managing multiple priorities in the past. "If we allocated resources according to these dot votes, for example, we'd put 50% of resources to growth, 25% to profit, and 25% to quality."

"Seems sensible," says Philippe. There is general agreement, and even enthusiasm from Sparks. Irie looks back at Sri, who seems like he's warming to the idea a bit.

"Then we're done!" pronounces Sparks. ■

5.2 Irie Discovers Shallow Alignment

"Hang on," says Philippe. "We did a quick vote with these dots, but I don't feel like that equals agreement on how to split effort between these objectives."

"Splitting is irrelevant," says Sparks dismissively. "Growth is top priority, so we do what's necessary there, like user access and AI. I also have a corporate partnership opportunity with VigGuard we need to fit in."

"That doesn't make sense," says Philippe. "Corporate deals are not profitable. We just agreed that we would consider profitability, even if it is a lower priority. I think these objectives mean we cannot serve the corporate market at all at this stage. We should focus on direct-to-consumer only."

"We are never going to get anywhere if we all interpret everything we come up with in completely different ways," Ella says in a quiet tone that nonetheless causes everyone to stop and listen. "We need to uncover these disagreements up front or we're not really aligned."

Irie says that her playbook has a list of techniques for uncovering hidden misunderstandings or disagreements. She explains, "You know that cartoon where everyone in the room says they agree, but each of them has a different geometric shape in their head? It's for those situations." ■

Figure 5-5. Agreement

Mining for Conflict

Have you ever thought everyone agreed to something, then had the plan fall apart when one key person reversed course after the meeting? Conflict can be hidden for many reasons, including an innocent misunderstanding or a hesitation to speak up in front of a crowd. Hidden misalignments can undermine any plan unless you bring them to light and deal with them early.

How to identify shallow alignment

If you observe any of these behaviors among your stakeholders, you may have achieved only *shallow* alignment, and it is likely worth probing further when you see situations like the following:

- When most stakeholders agree, but one or more of them are silent.
- Someone verbally agrees, but their body language suggests they are unhappy or have reservations.
- When asked whether they agree, someone uses hedging language like "I guess," "sure," or "whatever."
- Someone agrees in the room but communicates something different later.*
- Stakeholders cannot agree on priorities among objectives and cannot make trade-offs when objectives conflict.
- Someone uses vague words like "innovation," "synergy," or "impact" without including specific metrics, decisions, or deliverables.
- There are many objectives, but it is unclear how they come together or how overall success is measured.†
- Objectives or deliverables are agreed on, but no one commits to action items.
- Deliverables and dates are agreed on, but many are missed without explanation.

---

* This crosses the line from shallow alignment to fake alignment and is a strong indicator of a low-trust environment.

† The most common scenario is that each executive has their own objective related to the function they manage.

Conflict mining techniques

If you notice shallow alignment, you should verify that the team is really on the same page by deliberately probing for any remaining doubts or objections. Patrick Lencioni, author of *The Advantage* (Jossey-Bass, 2012), says that "one of the best ways for leaders to raise the level of healthy conflict on a team is by mining for conflict during meetings. This happens when they suspect that unearthed disagreement is lurking in the room and gently demand that people come clean."

Figure 5-6 lists some of our favorite conflict mining techniques, and Figure 5-7 provides more details about them

Figure 5-6. Some of our favorite conflict mining techniques

Open with a Bad Idea

Fist of Five

If/Thens

Challenge Round

Ask an Outsider

Ask for Commitment

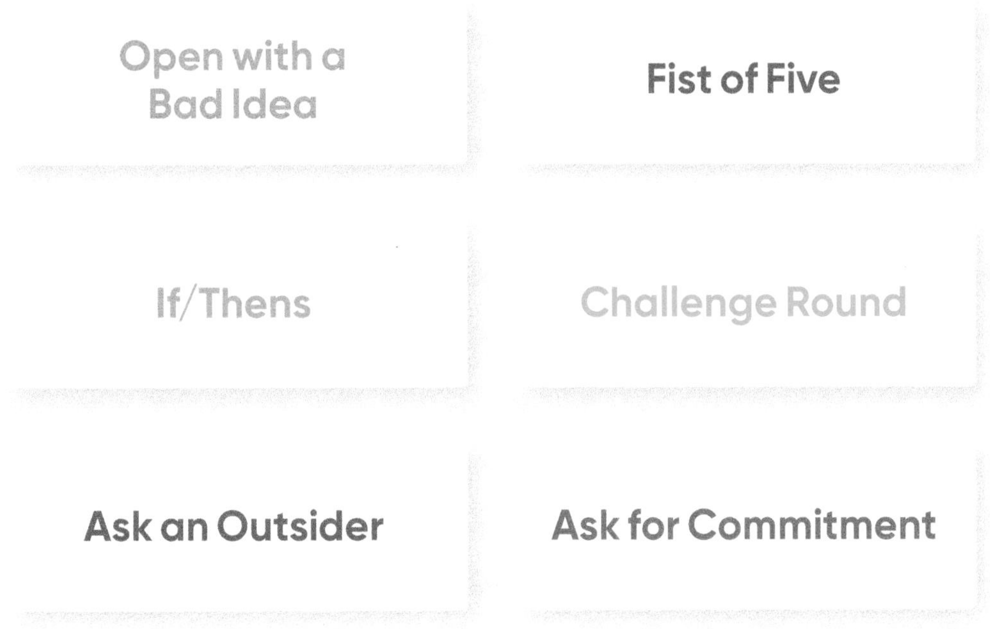

Figure 5-7. Conflict mining techniques

Open with a Bad Idea

Critique an idea constructively by introducing one with clear flaws. Discuss the idea and its flaws dispassionately and agree on why it is not the right thing to do.

When to use

Use to establish clear decision criteria and set an example of constructive ways to evaluate ideas.

Pro tips

Pick an example that has some good in it as well as a flaw that disqualifies it. Enumerate both the pros and cons. This clearly demonstrates that critical thinking and debate are safe things to do.

Fist of Five

Ask each participant to hold up a number of fingers between zero (no confidence) and five (complete confidence). Ask low scorers to articulate their doubts.

When to use

Use to assess how close to alignment a group really is before moving on, and to identify any remaining doubts.

Pro tips

Different cultures have different norms for how much confidence they will display. Over time, you will learn your team's defaults and how to interpret them.

If/Thens

Ask each participant to list the follow-on implications of the proposed course of action in the form of "if...then." For example,"If we add this item to the roadmap, then we need to drop something else."

When to use

Use to verify common understanding of the implications of a decision on work in flight, roadmaps, or changes to the organization.

Pro tips

Identify the "thens" that participants do not agree on for discussion and resolution. Capture the "thens" everyone agrees with as explicit decisions that also need follow-up.

Challenge Round

Ask each participant to articulate one reason a proposed course of action may fail. Discuss whether those risks effectively eliminate that option or whether there are acceptable mitigations.

When to use

Use to uncover hidden doubts among stakeholders when a tentative alignment has been reached but before asking for commitment.

Pro tips

Don't be tempted to lighten the mood by throwing out a ridiculous challenge like "What if the Earth is hit by an asteroid?" This may make people feel you are not looking for serious concerns.

Ask an Outsider

Ask a trusted person who has not been involved in the generation or evaluation of ideas or alternatives for advice on a tentative course of action before committing.

When to use

Use to avoid groupthink after your stakeholders have been deeply engaged. "Fresh eyes" may see problems the group failed to surface.

Pro tips

Ask someone who commands enough respect from your stakeholders to be taken seriously.

Ask for Commitment

Ask each participant individually if they are aligned to the decision and committed to making it work. If anyone says "no," or "I guess so," attempt to resolve their issues before asking again.

When to use

Use to bring out nagging doubts before finalizing a decision or direction.

Pro tips

Ask each person to commit to specific follow-up actions by specific dates to ensure the commitment is genuine.

Product Council

A "product council" is a small group of cross-functional stakeholders you bring together on a regular basis to make or review decisions that affect multiple teams. If a customer advisory board is your go-to set of advisors for what customers need from your product, then a product council is your group of advisors for your product strategy and cross-functional team processes. Regular product council meetings help you get to know your stakeholders better and surface disagreements early.

Here are some of the ways you can utilize a product council:

- Confirm a quarterly roadmap update.
- Debate potential shifts in organizational or product strategy.
- Determine how to respond to changes in your industry.
- Brainstorm solutions for cross-functional process problems.
- Establish or update objectives and key results for your product.
- Triage customer problems that could have technical or nontechnical solutions.
- Check in on a product launch or other cross-functional initiative.

You may have a leadership team meeting that includes your key stakeholders already. If not, it's helpful to assemble your own group. Monthly is common but you can meet more frequently if there are a lot of decisions to make.

Irie Mines for Conflict

"If we're not sure how to apply these objectives in making product decisions, why don't we write out our assumptions?" Irie suggests. "Write a sticky for each assumption in the form of 'if this, then that.' If we agree on something, we move on. If we disagree, we try to resolve it."

Sri's bright smile returns quickly. The group agrees and Irie creates a new section of the online whiteboard. They spend a few minutes silently writing their thoughts, one assumption per virtual sticky note. They group similar notes together for comparison, using affinity mapping.

"Well, at least we agree on a few things," Ella observes once the stickies are all grouped. They've marked groups of stickies with conflicting assumptions, and there are several key ones.

"How do you get to 55% growth?" Philippe asks Sparks.

"Yacob wanted 20% for quality," Sparks begins, but is cut off by Sri, who says, "He wanted 30."

"Sure, for a few weeks until we get it under control," replies Sparks.

"It's going to take longer than that!" says Sri.

"How long?" asks Sparks.

Figure 5-8. Irie's notes on mining for conflict

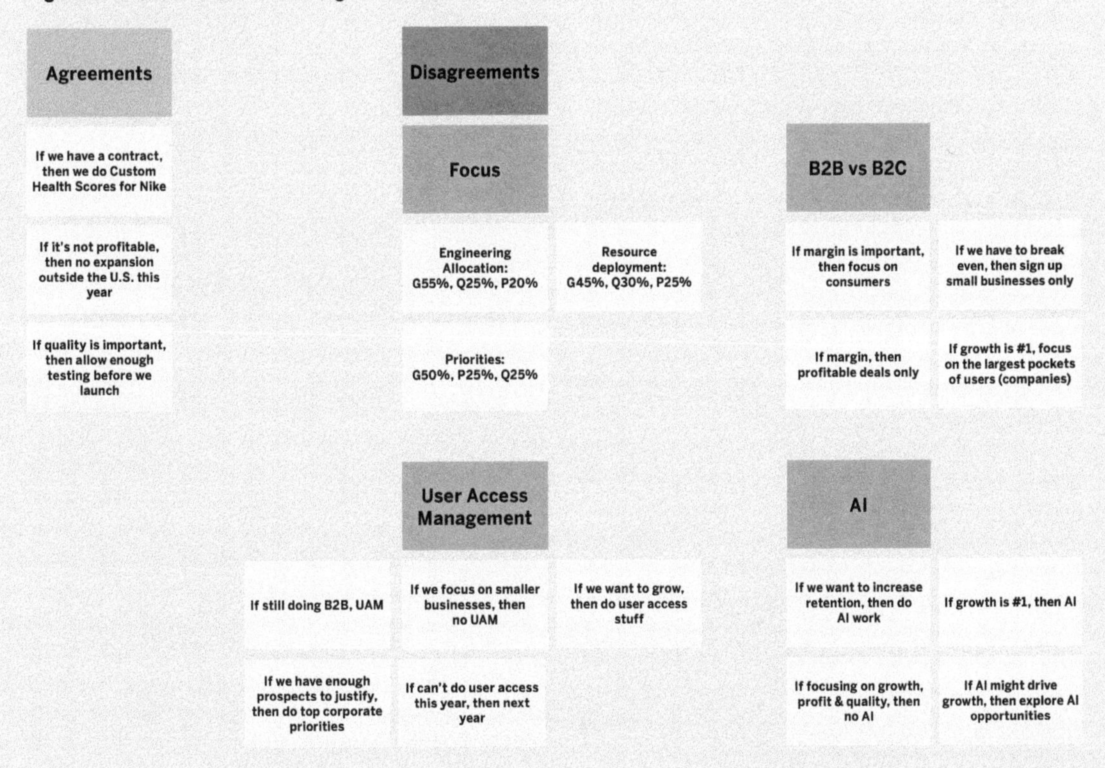

"I don't know!" says Sri, increasingly upset. "We won't know until we see support cases and outages go down."

"Cases are never going to go down while we are growing fast!" insists Sparks.

"Then we are growing too fast!" says Sri.

"How about we keep it at 30% for now," suggests Irie, reasonably, "and agree on an acceptable number of cases per customer per month. Once we get it down to that, we can reset the number to something lower like Yacob suggested."

"The investors gave us some benchmarks for things like that," volunteers Ella. "I can come back to this group with a proposed target."

"Fine," says Sparks. "30% for quality for now. We voted for 25%, so I say we take the missing 5% from margin, making that 20%."

There is general agreement to this, and Irie writes a note on the virtual whiteboard to record the decision.

Building on the momentum of a quick decision, Irie observes that the stickies related to the target market have a lot of overlap in them. "We all came at it from different angles, but I think we're saying that we're going to focus on consumers and small businesses because they are both profitable.

"I'm not saying that," says Sparks. "We said growth was the number one goal, and the fastest way to grow our user base is with large corporate deals where they sign up all their employees at once. Citi has over 70,000 employees in the US, and Arianna can get us all of those with one deal."

"We've been over—" Ella begins, but Philippe cuts her off.

"What if we use the rule of 40?" he asks. "Sorry, Ella. I am just excited. We keep going back and forth on growth versus profit and the rule of 40 addresses this directly."

Philippe explains that the rule of 40* says a SaaS company's growth rate and profit margin should add up to 40%. Each balances and limits the other. "We've been growing at just under 40% year over year the last two years. That means we need to have at least a little margin or we're not hitting the target investors look for."

"And that would make it hard to raise more money," says Sri, completing the thought.

"At the sort of valuation we'd all like, yes," confirms Philippe.

"But if we signed Citi we'd grow even faster," Sparks insists.

"Citi has a lot of people, yes, but alone they wouldn't justify the dent we'd put in margin," says Philippe. "I'll run the numbers to double-check, but I'm pretty sure that's what we'll see."

"It's not just Citi," says Sparks, exasperated.

"No, it's not just Citi," Irie interjects. "And that's just the thing. Every large enterprise customer has a different set of must-haves.

* For an overview, see Cody Slingerland, "What Is the Rule of 40?" CloudZero Blog, June 2023. *https://www.cloudzero.com/blog/rule-of-40*.

I've looked at the requests coming in from sales. We can't keep up."

"This is what I've been saying," adds Ella. "The reason these large deals are unprofitable is that we have to promise the moon to win them."

Irie continues. "I think we could build our way up the market to these larger deals over time, but I think we're talking years to cover everything they want."

"And in the meantime, we have a lot of easier wins," Ella adds.

Reluctantly, Sparks acknowledges the reality of the situation. "We need to give Arianna some guidance," he says. Irie is relieved to see Sparks conceding a point here. She is very glad she proposed this workshop, and she quickly agrees to work with Sergey to draft a target market definition with qualification questions sales can use.

Irie also returns to the virtual whiteboard and makes a note of their decision to use the rule of 40, which as long as they are below 40% top-line growth means they must break even on average with each new user.

"And I suppose this means we will not be doing user access management," says Sparks, appearing tired for the first time. Irie and Sri nod. Sparks asks if they can at least put it on the list of things to work up to, and the pair readily agree.

"Next you're going to tell me we can't do AI."

"Actually, I think you're right about that one, Sparks," Irie says. "From all the research, and conversations with the board, it's looking like we should do the AI work—over time, of course. I think it could be a differentiator and, importantly, it could be useful to customers in any segment."

"It would be splashy, too," Ella chimes in. "It could drive press and free trials."

"I'm not sure yet exactly what the best application is," says Irie, "but my team is already working on that and it's a lot easier for me to justify prioritizing it after this discussion."

"Who wrote 'no AI,' then?" asks Philippe.

Sri looks a bit embarrassed. "I just don't want to build a buzzword without knowing why we're doing it," he says. "But if we're all aligned that it fits our goals, then I am on board."

Irie adds another note to the board, reflecting their decision.

Figure 5-9. Irie's updated agreements

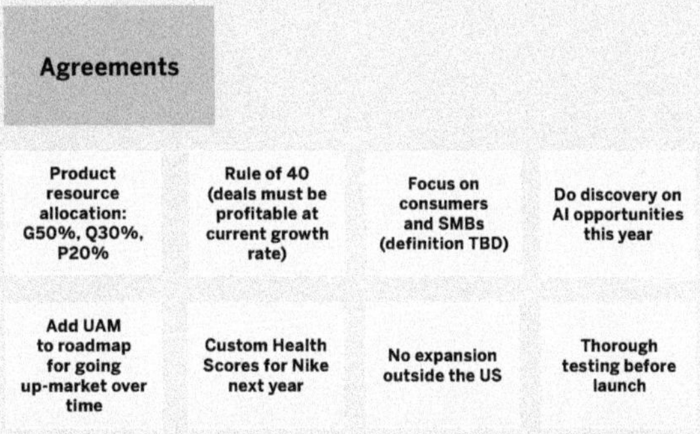

Agreements

Product resource allocation: G50%, Q30%, P20%	Rule of 40 (deals must be profitable at current growth rate)	Focus on consumers and SMBs (definition TBD)	Do discovery on AI opportunities this year
Add UAM to roadmap for going up-market over time	Custom Health Scores for Nike next year	No expansion outside the US	Thorough testing before launch

She asks if they all agree that these are their decisions. Ella and Philippe nod but no one speaks for a moment. "Let's do Fist of Five," Irie says, and explains the ritual.

Ella holds up five fingers, explaining, "I'm just glad we're having this conversation. We've never done this right." She thanks Irie for bringing them together.

Philippe and Sri hold up four fingers each, agreeing that they are waiting for final details from Ella and Irie on a few things.

Sparks holds up three fingers. "This is fine for now," he says, "but the situation may…evolve and we may have to make some changes."

"What's that supposed to mean?" asks Sri.

Sparks says that he can't talk about specifics, but if at some point there is a significant change in their business like a partnership or a new round of investment, that might change their objectives.

Sri points out that this would always be true, and asks if he brings it up now for a specific reason. Sparks says no, but he just wanted to be clear.

"Clarity is good," Irie says. "Can we all agree we'll use these points for product decisions until such an event?"

"That might happen next month or never," says Sri. "Why don't we get this group together every month or so to review and update the goals?"

"I've done this before and it's really helpful," observes Irie. "We called the group a 'product council.' We met once a month to review the roadmap, but more frequently if there were big changes happening. Right now there is a lot of work to do to put together product objectives and the roadmap. I want you all to be involved since product decisions affect all of us. For now, we could meet every week or two while we work out the product objectives and put together the roadmap. Then we can reduce the frequency as we get into a rhythm. Does that work?" Everyone agrees. Even Sparks gives an acquiescing nod. ■

5.3 Irie Aligns on Product Objectives

The following week, the product council meets officially for the first time. Sri is remote because he has a medical appointment later in the day. "My shoulder has been killing me since I pulled something in a basketball game last weekend," he explains. After giving him some sympathy, Irie introduces the topic for the meeting: developing product objectives.

Sparks looks up at this. "I thought we just did the product objectives last week," he says.

"We worked out which organizational objectives drive product strategy," Irie says. "It's a good start, but product objectives need to be specific to things the team can impact short-term." She pulls up a slide from her playbook entitled "Aligning Product Objectives to Organizational Objectives," and shares it with the room.

"I assume you can easily split off things like bug fixes from new feature work," says Ella, "and I think that sounds smart, but how do we allocate 'resources' to growth versus margin?"

Irie explains that product objectives work best when they are tied to customer behavior that can be measured frequently. "You can only see changes in growth and profit over many quarters," she explains, "and there are a lot of other variables that might affect these results, like marketing campaigns, or a big deal coming in. That makes it hard for Product Teams to know what effect they have when they make changes, like adding features or updating the UI. So Product Teams measure things like conversion rate as a proxy for growth, and things like usage as stand-ins for margin."

"Wait," says Ella, "can you connect those dots for me? How does usage affect margin? Does more usage drive up hosting costs or something?"

"I can explain that," Philippe volunteers. "Churn is when we lose a customer. We then have to spend money to acquire a new customer to replace them, just to stay even. It drives up the cost of growth to have a leaky bucket like that," he explains.

"That's churn," says Ella. "What about usage?"

"Churn happens slowly," explains Irie. "It can take months for us to realize someone has churned. So we want to measure something that changes more quickly. Usage is generally correlated with sticking around—the opposite of churn."

"That makes sense," says Sri. "I use an app to track my workouts nearly every day. It costs me a hundred bucks a year, but there is no way I'm giving it up."

"The longer you stay," Philippe says, "the more profitable you are for that company." ■

Derive Product Objectives from Organizational Objectives

Executives tend to think in terms of broad goals, like revenue, growth, and profit. They may also set goals related to specific strategies, like market expansion or technical initiatives (like AI). But these goals are often too generic to be useful in guiding the day-to-day work of a Product Team. Product Teams work best when they have objectives that can be measured quickly and frequently (leading indicators), whereas goals at the organizational level tend to be slow moving (lagging indicators).

Some teams may think that launching a feature is an obvious candidate for a product objective, but if your goal is "ship the AI feature," how do you know whether or not that feature had a positive impact on corporate goals like revenue, growth, or profit? We select and design features to create beneficial changes for our customers and our business, so product objectives should describe and measure those hoped-for changes.

Defining the objective as the *change* you seek rather than the *feature* you're building empowers the Product Team to adjust the scope and details of feature work over time to meet the objective. In some cases, it can even justify abandoning a planned feature completely if testing reveals that some other approach is more likely to generate the desired outcome.*

Former VP of product for Netflix, Gibson Biddle, explains that their number one objective was retention. Retention was crucial to maintaining their rapid growth, but because retention is hard to change and moves slowly, Gibson's Product Teams focused on various ways of measuring engagement. The more people watched, the more likely they were to keep watching and keep paying. Netflix's roadmaps were filled with initiatives and experiments designed to drive engagement, and their success metrics were things like "Percent of members who stream at least 15 minutes of video in a month" and "Percent of new members who rate at least 50 movies in their first six weeks with the service."

To effectively create product objectives from organizational objectives, then, we must ask these questions: 1) Which organizational objectives

* See *Product Roadmaps Relaunched* (O'Reilly 2017) for more on how to make outcome-based roadmaps based on organizational objectives.

can the product contribute to directly? and 2) What proxy metrics can the product affect that will drive organizational objectives?

Which organizational objectives can the product contribute to directly?

A Product Team can contribute *indirectly* to many organizational objectives. They can watch expenditures to improve cash flow, practice constructive feedback to help build a great culture, and get on calls to help sales close deals.

Product Teams can also add the features that drive sales and retention by delighting customers with the capabilities they want and need. They can make products easier to set up, learn, and use to improve customers' ability to gain value from the products, which can drive satisfaction and retention.

But if sales and retention move slowly and have many other contributors, how can we measure the Product Team's contribution? Good product objectives involve changes in customer behavior, customer value, or business value that the Product Team can affect directly and measurably.

What proxy metrics can the product affect?

In a famous example, Facebook found that users who add at least seven friends in their first ten days on the app tended to be more engaged and stay with the product longer. This led them to set an objective of getting more new users to add more friends within that time frame. The Product Team then tested various changes to the onboarding experience to encourage adding friends. In the end, the causal relationship between more friends and increased retention (and therefore faster growth) was validated, and the company still tracks similar metrics.*

This is a classic example of a *proxy metric*, a metric that evidence suggests has a causal relationship with the metric we care about but is more useful because it is easier to measure, easier to isolate from other influences, or changes more quickly. In his book *Outcomes over Output* (Sense & Respond Press, 2019), Joshua Seiden ties the proxy metric (product objective) with

* A concise video on this from Chamath Palihapitiya at Meta: *http://youtu.be/raIUQP71SBU.*

the organizational objective by asking, "What are the customer behaviors that drive business results?"

Facebook sets product objectives as changes to customer behavior, as Joshua suggests, but another approach to proxy metrics is to measure the value customers create for themselves via the product. This might be the completion of workflows or tasks. For example, a marketing campaign management tool might measure the number of campaigns launched.

Even better, your product objective might be tied to the customer's results. Omnichannel retail platform provider NewStore measures both the customer revenue generated through their platform and the increases in that revenue driven by the use of their tools. NewStore CPO Marcus Bittrich says, "By focusing our efforts on making the customer successful, we drive our own business forward."

Proxy metrics are the typical method for measuring the value generated by product changes, but business value can sometimes be measured directly. If your customers pay for usage, e.g., by the minute or the transaction, then increasing minutes or transactions increases revenue directly. Increasing in-app purchases can also be driven directly by changes in the product.

Whether your product objectives measure customer behavior, customer value, or direct business value, it's critical to identify and align with your stakeholders on product objectives that are derived from organizational objectives.*

* For a deep dive on setting your product's metrics and aligning them with your growth strategy, read Matt Lerner's *Growth Levers and How to Find Them* (SYSTM, 2023).

Irie Establishes Product Objectives

At the next meeting of her new product council, this time in person, Irie summarizes the three organizational objectives they've agreed to on the whiteboard:

Figure 5-10. New Helthex OKRs

Growth 50%	Profit 30%	Quality 20%

She suggests that they establish some proxy metrics for the technology team to own.

Sri agrees, saying, "If we can directly tie them to the company OKRs, I think everyone will see the value in that."

Ella points out that they already agreed that they'd revisit the quality problem if customer cases related to quality went down.

"Right, so we'd want to see a reduction in customer reports of high-severity bugs, responsiveness, or availability problems before making a change," suggests Irie.

"We could measure these per 1,000 customers," Sri suggests. "Otherwise, as Sparks suggests, the raw number doesn't account for organic growth."

Sparks suggests they go back to the company OKRs. "App Store ratings, NPS, and CSat all seem like things that the product could drive. There's also Custom Health Scores for Nike," he adds.

Ella explains to Irie that Custom Health Scores was written into the Nike contract. "We have to do it at some point," she says, "but they've delayed their implementation until next year for internal reasons, so it's not a hard requirement for this year anymore."

Irie asks about Sparks's other suggestions. "They all seem like ways of measuring customer sentiment."

"It seems important and, when we did the OKRs, everyone had their own ideas on how to measure it," Sri admits.

"All three of these feel like proxy measures," Irie observes. "But proxies for what? What are we really hoping will change if customers are happier? Have we got a problem with churn?"

"It's higher than we'd like," Ella says. "90% retention is good for a SaaS business, but we hover just under 80%." Irie makes a note that they'd like to increase retention.

"Less churn and more retention supports both margin and growth, right?" asks Sri.

"Yes," volunteers Philippe, "but it drives margin more directly by reducing the cost of replacing lost customers."*

"So it's a proxy for margin?" asks Sri.

"Yes, but it still moves pretty slowly," Irie explains, "I've treated it that way before and I usually need something faster-moving that correlates with retention. Engagement

* See Saravana Kumar's piece, "Customer Retention Versus Customer Acquisition," *Forbes*, December 2022, *https://www.forbes.com/sites/forbesbusinesscouncil/2022/12/12/customer-retention-versus-customer-acquisition*.

Figure 5-11. Helthex organizational objectives and product objectives

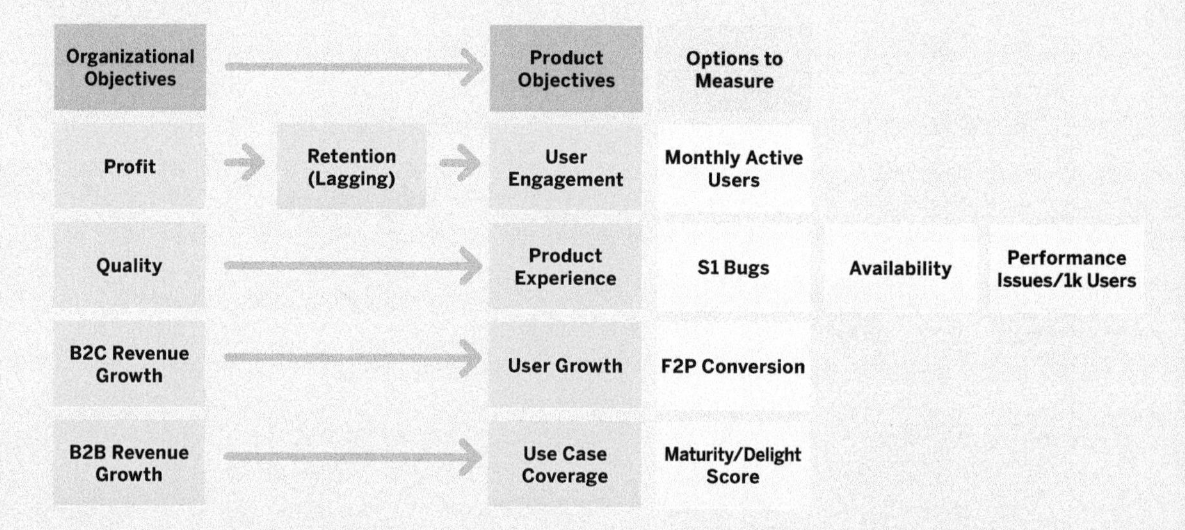

metrics like how frequently people log in are better leading indicators. They are easier to make an impact on quickly, and generally, people stay with things they use a lot."

"So engagement leads to retention, which leads to margin. That makes sense," says Sri. "He then puts those two words on the board next to Profit and connects them with arrows.

Sri then asks, "Would app responsiveness also feed into retention?"

"That's a testable hypothesis," Irie says. "Do people who churn complain about the app being unresponsive?"

Ella admits that she's heard complaints about slowness but she doesn't know how it correlates with churn. "I can ask Liandri to look into this," she adds.

"What sort of proxy would you want for revenue growth?" asks Sparks, starting to get into the conversation.

"For a B2C app like ours, we should be looking at conversion from free to paid and maybe user growth, or churn by cohort," Irie replies quickly.

"But we have corporate clients, too," Sparks points out.

"Yeah," Irie replies. "Since it's a small number of large deals and the company pays for their users, we can't use conversion or user growth."

"The ultimate measure would be win rate," says Ella.

"Yes, but that definitely lags," says Philippe. "Our sales cycle is 12–18 months. Could it be like how many 'yes' answers we can give in an RFP?"

Irie is thinking hard. "Maybe, sort of," she says. "If we know what the corporate market really requires, I think it could come down to filling out the necessary use cases."

"How are use cases a metric?" asks Sparks.

"Sorry, I should explain," says Irie. "A 'use case' or 'job to be done' is something a user wants to accomplish. It can be as simple as a task, like answering an email, or a little more general like passing the time while waiting in line for coffee. Each involves interacting with multiple parts of the product to accomplish what the user wants."

Ella chimes in, "Liandri tells me our corporate users want to do things like review usage by their employees, and target people at risk for certain health conditions with messaging."

"We would do research with users to define these core use cases," Irie explains. "Then we'd score them as to how completely we cover them in the app. We could even score them as to how delightful the experience is. We can evaluate this ourselves, but it would be better to get feedback from customers and prospects and turn their feedback into a score." She adds, "And then the objective is to increase that score over time."

Sparks seems to be absorbing what she is saying, and Irie updates the whiteboard.

The group seems quiet now, so Irie leads them through another round of mining for conflict. They end up agreeing that what they've come up with is good, but there are still too many product objectives and they have no specific numbers.

"I'd need to see the specific numbers for all of these metrics so we can agree what is achievable," says Philippe after a moment. Sri seconds this. The group agrees that they'll meet again in two weeks to review specific numbers, pick a smaller number of product objectives to pursue, and assign owners to bring proposals for each. Irie makes a bulleted list of action items on the whiteboard:

- Philippe will work with José and Christina on modeling user engagement, growth, conversion, and churn.
- Ella and Sri will work with Liandri on customer escalation numbers.
- Irie will work with Arianna to propose a way to measure use case coverage.

"These goals are fine," says Sparks. "But what are we actually going to ship? And where does AI fit in?"

Everyone looks to Irie. "If we can refine these objectives," she says, "we'll have enough of a foundation to work on a roadmap."

"And I see AI all over these objectives," Sri says. The others wait for him to explain, and he says, "If we can make the app more interactive, it should improve monthly active users, growth, conversion, and churn, maybe even delight scores if we do it well."

"That's great," says Philippe. "So let's think of AI as our number one experiment in driving up user engagement. If it works, it should improve things across the board."

"So AI is our number one priority," says Sparks, crossing his arms. "What a shock."

"No," says Philippe. "User engagement is our number one priority. AI is our best guess right now as to how to accomplish that." ■

5.4 Irie Co-develops a Product Roadmap

Two weeks later, the product council regroups, having accomplished most of their action items from the previous meeting.

"Making the roadmap should be simple," says Sparks. "We rank all the work based on these objectives and do it as fast as we can."

"Hang on," says Sri, "We need to investigate feasibility. I think there is architecture work we need to do before some of these things will be possible."

"And maybe we should look for some quick wins to help accelerate growth," adds Ella.

"That makes sense," Philippe says. "Sometimes the smallest things can have huge ROI."

"So we could rank things by impact on objectives versus level of effort," Sri says. "Then we'd know which are the quick wins, which are worth the wait—"

"And which aren't worth the effort," Philippe says, finishing Sri's thought.

Irie nods along with what everyone is saying, smiling at their level of enthusiasm and their cooperation in the process. The group finally turns to her and Sri asks, "So how do we go about making a roadmap?"

"I'm glad you asked," she replies. "You all are describing ways to prioritize features, which is great. I have a formula I've used that I think will work with our objectives. But that's only a small part of a product roadmap," Irie continues. "A proper roadmap begins with a strong vision of the value we hope to bring to the world."

"We can start with Liz's original vision statement," Sparks volunteers.

"That's right," says Irie. "The roadmap must then articulate the customer problems we need to solve to achieve that vision. Finally, we use our objectives as a lens to identify the most important of those problems for us to focus on."

"That sounds a bit hand-wavy," Ella says. "Are you saying the roadmap shouldn't have features on it?" Sparks and Sri turn toward Irie, waiting for her answer.

Irie is pleased that the group now trusts her enough to look to her for these answers "No, features are fine," she replies, "once we are confident they solve the problem effectively and are feasible and profitable."

"I like the sound of that," volunteers Philippe, and Sri seconds.

"So, what, you have to do months of research before you can commit to anything on the roadmap?" asks Sparks, incredulous.

"Sometimes, yes," interjects Sri. "Right now I can't commit to a date for user access management or for any of the AI features, because there are too many unknowns."

"And we often need to research customer problems to be sure we're working on the things that will make a difference," adds Irie, "and then validate that our proposed solutions actually deliver for the customer."

Irie sees Sparks and Ella are ready to object, so she quickly adds, "But we've actually done a lot of that work already. Alex has given us a good base of insight into customer problems, and José has done a lot of prototype testing covering several of the proposals on our list. I'm more worried about the feasibility side of UAM and AI," she adds. "How much time would you need to do that research, Sri?" she asks.

Sri pauses, absorbing that Irie is now challenging him a bit. "I could have a better handle on the scope of UAM in two weeks. The AI thing has a lot more uncertainty. A couple months would give us a rough idea."

Ella asks if they could be working on quick wins in the meantime. "Things you already know the scope of?"

"Yes, we would ideally start working on things that are already pretty well-known," says Irie. "That is the process I've used before. We researched new things in parallel with implementing things we'd already researched. We did put features on the roadmap when we were confident they would solve a customer problem and help us achieve our objectives. And we used 'themes' to describe customer problems where we were still figuring out the best approach."

Ella and the rest look quizzical, so Irie opens up her playbook once again to dive into roadmaps. ■

What Is a Product Roadmap?

Most people think of a roadmap as a set of features and committed dates for shipping them. That's actually closer to the definition of an implementation plan, release plan, or project plan. A good roadmap may have dates (or broad time frames such as quarters) and it may have features, but it's actually much more than that.

A good roadmap is a statement of strategy that helps stakeholders understand the destination and the obstacles along the way. It also provides guidance for navigating those obstacles without prescribing an exact route. It's more like a GPS that recalculates as needed than it is a fixed set of turn-by-turn directions.

This strategic perspective focuses on the product objectives, customer problems, and organizational goals, instead of output like launching features. This is important because knowing the desired result gives the teams doing the work the necessary context to make tactical decisions along the way. It may even provide justification for abandoning a planned feature if research shows it will not solve an important customer problem in a profitable way.

This approach to an "outcomes" roadmap is described in depth in *Product Roadmaps Relaunched* (O'Reilly 2017), coauthored by Bruce. The Wombatter product roadmap from the book (Figure 5-12) showcases examples of each of the key components of a good product roadmap.

These five components—a product vision, timeframes, themes, organizational objectives, and a disclaimer—form the core of a good product roadmap. Together they provide a picture of the value the product intends to deliver, in a way that stakeholders and customers can understand (Figure 5-13).

Figure 5-12. Sample product roadmap—The Wombatter Hose

THE WOMBATTER Hose

(1) **PRODUCT VISION**

Perfecting American lawns and landscapes by perfecting water delivery

H1'25	H2'25	2026	Future
Indestructible Hose Objectives: • Increase unit sales • Decrease number of returns • Decrease overall defects (4)	**Delicate Flower Management** Objective: • Double ASP	**Putting Green Evenness for Lawns**	**Infinite Extensibility**
	Severe Weather Handling Objective: • NE Expansion	**Extended Reach**	**Fertilizer Delivery**

(2) (3)

(5) Updated 6/30/24, subject to change.

1. **Product Vision**
2. **Timeframes**
3. **Themes**
4. **Organizational Objectives**
5. **Disclaimer**

Figure 5-13. Components of a good product roadmap

Component	Description	Importance
Product vision	A concise description of how your customer will benefit from your product	By focusing on customer outcomes, a product vision provides guidance to Product Teams without dictating specific designs, specs, or features.
Timeframes	Broad blocks of time like calendar quarters or even the popular dateless "Now, Next, Later" format*	Precise date commitments divert attention from themes and objectives and do not allow for the uncertain and iterative nature of innovation inherent in new product development.
Themes	Customer problems to be solved to achieve the product vision	Focusing on customer value helps teams evaluate possible solutions and be accountable for results (outcomes) rather than simply delivery (outputs).
Objectives	What will be measurably different for your org with a successful product	Focusing on the value to the organization helps evaluate the business viability of solutions and prioritize those with the highest ROI.
Disclaimer	A clear statement that the roadmap is subject to change	A disclaimer makes clear that the roadmap is always evolving and protects your stakeholders from mistaking it as a set of committed deliverables and dates.

A great roadmap is a living document used to communicate with your stakeholders. It should be as visible as possible, so that your stakeholders are always informed and never surprised about what your team is working on. Roadmaps are meant to be referred to and discussed often. There are a few practices worth considering:

- Present your updated roadmap at every all-hands meeting.
- Post your roadmap on an internal wiki or shared drive.
- Publish a public version of your roadmap on your website (perhaps with less detail than your internal version).
- Share a short video of you presenting your roadmap.
- Post an enlarged roadmap in the break room or work area.
- Provide a printable PDF so people can keep the roadmap at their desk.

---

* Janna Bastow, CEO of ProdPad, popularized the "Now, Next, Later" roadmap, which contains no dates at all. We favor this timeframe format in many circumstances because it focuses on value rather than delivery dates.

Irie Proposes a Helthex Roadmap

"Do we really need all of this?" asks Sparks. "Isn't the vision obvious?"

"I don't think so," says Sri. "Don't get me wrong," he adds quickly. "Liz does a great job of painting a picture of a world with 'personalized medical advice for anyone, anytime, anywhere.' But when we are arguing about this or that feature or shifting from one integration to another, I get a little lost."

"I'm getting the same thing from my team," volunteers Irie. "And José and the rest of the design team feel like their user research is ignored."

"If we're supposed to be making a better world for the users, that doesn't seem right," Ella says. "And having clear themes would make it easy for Sergey's team to write benefit statements. It would actually help us get ahead of releases with our marketing messaging even if we don't know the exact details of every feature."

Sparks rolls his eyes but agrees they should proceed.

The next time the group gets together, Irie leads them through developing the core components of a roadmap. She has them use silent writing, affinity mapping, dot voting, and other workshopping techniques to drive collaboration and alignment.

After four sessions, and multiple iterations, the product council eventually aligns on a simple roadmap that organizes themes according to a smaller set of product objectives and specific metrics they hope to achieve.

Figure 5-14. The Helthex Roadmap

HELTHEX Roadmap

Personalized medical advice for anyone, anytime, anywhere.

Product Objectives (Organizational Objectives)	Metric	Now	Next	Considering
Engagement (Margin)	Monthly Active Users (MAU)	Personalized recommendations	Proactive recommendations	Accountability sharing
Experience (Quality)	S1 issues per 1k users	Handle traffic spikes gracefully (Q1)	Improve perceived load time on key flows	<6 min of unplanned downtime (Five nines)
B2C Conversion (Growth)	F2P conversion	Interactive experience	Personalized content	Symptom diagnosis
B2B Coverage (Growth)	Use case coverage	Admin user controls (Q1)	Data portability (Q2)	Privacy compliance

Subject to change. Uncertainty increases as timeframes get further away.

The initial draft includes no dates. Sparks objects, asking for "commitments" for all major initiatives. Sri reiterates his objection to premature date commitments.

"Well, then this roadmap is meaningless," says Sparks with a dismissive tone. "It might as well say 'we'll deliver some stuff at some point.'"

"Actually," Irie interjects, "I agree with you both."

Sparks and Sri look surprised and expectant, so Irie continues. "We can't commit to dates on things we haven't had time to investigate well enough, but we can have a release plan for things where we've done that work."

Sri agrees, saying that they have reasonable projections now for handling traffic spikes and user access management. "And actually," continues Sri, "Eitan and Wei have done a great job on looking into data portability. I think we know the rough scope for that too."

Irie suggests they add a calendar quarter for a projected ship date to things they are comfortable with, leaving the ones that require more investigation without dates until they are more certain.

"Like quick wins," adds Ella to general approval.

Irie then suggests they bring someone else in to provide feedback on their work. "I'd like to show our working draft to Liz," she says, "and see what she thinks."

"We need to give her some space to get better," says Sparks. "How about Arianna? She can tell us whether this will fly with customers."

"Or Sergey," suggests Ella. "He'll bring the analytical perspective."

At Philippe's suggestion, they ask those stakeholders to join their next session. Irie walks them through the roadmap and invites them to comment and ask questions.

"I like how the roadmap is grounded in objectives we want to achieve," Sergey says. "It is like a marketing campaign. You have a number for qualified leads you must hit and you run ads, send emails, and tweak offers until you hit your target."

"I don't like how unclear these themes are," says Arianna. "'Interactive experience' could be anything, no? Why don't we just say clearly 'AI chatbot' if that is what we mean to do?"

"We are investigating an AI bot to meet the need for more personalized recommendations," Irie explains. "And our preliminary research shows people respond well to live chat with a human, but we don't know yet whether a chatbot is a good substitute."

"Or whether it will even work," says Philippe. "I've seen some truly terrible automated chat agents."

"Alvex has the cutting-edge tech here," says Sparks. "We just need to get something working. That'll get people excited and we can refine it later."

"I for one am happy that we do not commit to a solution before we test it," says Sergey, crossing his arms. "Customers want interactivity. Fine. Maybe the best way is Liandri hires two hundred nurses from South Africa to answer chats."

"A large service organization isn't scalable," says Philippe.

"And it won't help our valuation," says Sparks.

"Let's not debate the solution right now," says Irie calmly. "My point was that the solution is, in fact, debatable. Until we can confidently commit to a specific solution, the roadmap should keep us focused on the problem, not the solution."

Irie continues. "Arianna, you remember the conversation we had with RightBank?"

"Yes, you were great with them," she replies.

"If I presented this roadmap to them without the metrics," Irie continues, "I would be happy to say to them that we are investigating AI chatbots and other solutions to see what feedback they have. I could even offer to let them test prototypes to ensure they fit their expectations. Do you think that would satisfy them?"

"I think they would love to be involved," says Arianna, excited. "I can set something up for next week!" ■

Takeaways

Like Irie, you will likely face multiple conflicting demands, priorities, and objectives coming in from around your organization. Aligning with stakeholders on organizational objectives will make it easier to define product objectives and make product decisions.

- Use workshopping to get your stakeholders on the same page. Techniques like silent writing, affinity mapping, and dot voting can help your group come up with new ideas and focus on the most important ones. Workshops are particularly effective to align on organizational and product objectives.
- Mining for conflict helps you identify shallow alignment among your stakeholders. Try opening with a bad idea to demonstrate constructive critique, or use a challenge round to force everyone to give a reason why the decision might fail.
- Create a cross-functional product council to ensure continued alignment among your varied stakeholders.
- Derive product objectives from organizational objectives by thinking about how the product can directly contribute. Consider starting with proxy metrics that are easier and quicker to measure.
- Create a product roadmap based on themes or problems to solve, not product features. This gives the Product Team more flexibility in how they achieve the product objectives.

As Irie will soon discover, an aligned product roadmap is also a great defense against shifting priorities or late-breaking requests.

Align with stakeholders on when to say "no" together

Chapter 6

Changes

After finally aligning everyone around a roadmap, you may think the task is complete. Very often, however, maintaining alignment on a roadmap over time is actually more difficult than establishing it in the first place.

Most roadmaps require not only longer-term strategic changes, which can be planned ahead of time, but also urgent changes to the current priorities, which require trade-offs and difficult decision making. This chapter will build on the process we discuss in Chapter 5, by showing you how to work with stakeholders as things change.

In this chapter you will learn how to do the following:

- Manage ongoing requests.
- Plan for routine roadmap updates, and avoid Alignment Decay.
- Decide whether to say yes or no to a roadmap update.
- Say "no" tactfully and effectively.

Let's check in with Irie six weeks into the new roadmap. Engineering and data science are making progress on the new AI features. A few features have already been released in the app. But some stakeholders still seem confused about how they can contribute to the roadmap when they have new ideas.

6.1 Irie Reduces Distractions

Irie hears a female voice speaking loudly as she approaches the conference room for the weekly joint product and design meeting. Entering, she hears Min, the product manager for the recommendation engine, speaking to Christina. "He just won't leave us alone!" she says.

Christina echoes Min's concern. "Plus he's constantly changing his mind."

"Who are we talking about?" asks Irie.

"Sparks," explains Min. "He's making requests every day. He's trying to change my roadmap. He's pestering my engineers!"

"And that's on top of all the salespeople who are still asking us for things every day," adds Christina. "Weren't you going to work on an intake process for feature requests?"

"I've been so busy trying to get the roadmap review meeting together, I forgot about it," Irie says. "Sounds like it can't wait."

Eitan joins the meeting, his oversized image appearing pixelated on the wall-mounted monitor. He confirms what the others have said. "It's not just Sparks and sales, though," he explains. "Support Slacks me about things whenever they run into an issue. I don't blame them, but the interruptions make it hard to concentrate."

José walks in the room, and some design team members join online. They also confirm that they are constantly bombarded with requests from various stakeholders.

"In the past I've used an intake process that works pretty well for features, bugs, and even support of individual deals," Irie says. "Why don't I walk you all through what we did and we can develop a proposal for a similar process." ■

Intake Process

If you or your team are fielding constant requests from all directions, it can be overwhelming. Using an intake process to consolidate and periodically review those requests can dramatically reduce the context switching, constant messages, and distractions that can slow down your team's day-to-day work. It also helps your stakeholders understand their role in defining or refining the product.

The goal of the intake process is to decide whether to work on each request. Throughout the intake process, you'll work with stakeholders to gain alignment, but ultimately you must make decisions based on what is best for the customers and for the business.

You can accept submissions from anyone who has an opinion about the product, including internal stakeholders, external partners, and customers. You can use specialized idea submission tools or something as simple as Google Forms to consolidate the requests and look for patterns.* For customer requests, you might consider having a team like customer support analyze submissions and bring you the top requests.

The intake process should have a few key steps:

- **Step 1: Submit requests**—Stakeholders submit requests using your chosen intake tool.
- **Step 2: Triage requests**—Review and triage requests periodically, typically every week or two, to quickly categorize requests and prepare for follow-up.
- **Step 3: Follow up**—Follow up with stakeholders for more information, discuss prioritization and trade-offs, then re-triage the requests.
- **Step 4: Decide on a plan**—Decide on next steps for each item and update the status.

When you roll out the process, explain these steps to your stakeholders. You should also share the dates or cadence of the triage meetings to set expectations about when you will get back to them with a response. Next, we'll detail each of these steps.

---

* You can find suggested tools for consolidating stakeholder ideas on our website: *alignedthebook.com*.

Step 1: Submit requests

The purpose of formalizing the ideas submission process is not only to stay organized but also to set proper expectations with stakeholders about their collaborative role in defining and evaluating ideas. Explain to stakeholders that answering the questions on the submission form is critical to evaluate requests. These questions are not a roadblock or a delay tactic—they are the basis for the prioritization process.

Figure 6-1 shows a list of questions you might put on an input form for internal stakeholders. If stakeholders have trouble with the questions, you can help them work through the form, making it easier for them to answer the questions on their own next time.

Keep it simple. If the form is too cumbersome to fill in, they may give up on the intake process and come to you directly anyway. On the other hand, making the process too easy might encourage people to submit requests without thinking them through first, so you have to balance ease of use with thoroughness.

If you are having trouble gaining adoption of the intake process, you may want to start with fewer questions, such as "What is the problem?", "Who is it for?", "What is the value of solving the problem?", and "What is your solution idea?" This will give you a sense of the request, and then you can follow up directly to answer the remaining questions.

Figure 6-1. Intake form for product ideas—sample questions

	Example question	Reason
Why	What is the problem you're trying to solve?	Defining the problem gets them thinking about why they want this thing. If they don't currently have the requested feature, be prepared to follow up.
	What is the value to the customer and/or to the business of solving this problem? How did you calculate that?	Focusing on the value of solving the problem communicates that "nice to have" features are likely to be deprioritized.
	Which company goal or strategic initiative does this request align with?	Encourages the requestor to think about the relevance of this request versus the aligned strategy and goals.
Who	Who is the request for? Which specific customer or internal team has the problem described?	It is helpful for the requestor to articulate who the requested feature is for, to avoid confusion. For example, maybe you thought it was for a customer, but actually it's for an internal team.
How Often/ How Many	How often does the problem happen? How many customers have this problem (if it's a customer problem)?	This gets at how pervasive the problem is, which can help you understand the scale of the problem.
How Urgent	How urgent is it to solve this problem? What is the cost of delay?* What is the impact if we don't work on this right now?	In addition to value, level of urgency can be an important factor in determining prioritization and timing of handling requests.
How	What else have you tried to solve the problem? Is there a workaround? How easy or difficult is the workaround?	This question gets them to think about what else they could do to solve the problem, like a manual version of the solution.
What	Share any screenshots or videos you have of the problem.	Can be helpful if the words alone don't clearly communicate the problem.
Specific Ideas	Do you have any solution ideas? If so, what are they?	Waiting until the end to ask about the solution can help people think through the problem and be more articulate in how they frame the request.

* Cost of delay is a common formula for estimating the impact of not doing something proposed. For example, if a feature is projected to increase the average sale price (ASP) by $10 and unit sales are typically 100,000 per month, then the cost of delaying that feature is $1 million per month.

Step 2: Triage requests

The goal of triaging requests is to decide what to work on now, later, or not at all. Sometimes that decision is straightforward and sometimes it's more nuanced. You should be able to categorize each request either immediately or after some follow up to gather information missing from the submission form. You can triage individually, or as a group of PMs, depending on the scale and interdependence of your products.

Triage categories should run from "yes" to "definitely not" and can have several intermediary categories depending on your stakeholders and your organizational culture. For example, if you have a senior stakeholder who simply will not accept a "no" answer, then you might want to call the category "not in the foreseeable future," which is essentially the same thing but might be a better way for you to move forward.

When triaging requests, first translate them into problems to be solved. Then you can sort the requests into categories like these:

- **Yes (will consider now):** These are problems that unquestionably need to be considered now (even if they are scheduled later), such as compliance with a new regulation that will affect how your customers use your product.
- **Yes (will consider soon):** These are problems that are less urgent, but still important* and should be considered soon, ideally within the next few months or the next time you do a major update to your roadmap.
- **No (not in the foreseeable future):** These are problems that are not aligned with your objectives or strategy, or are not feasible at this time. You can revisit these later, but for now they are not going on your list.
- **No (won't do):** These are requests that are illegal, unethical, will harm customers, or are simply impossible, and they are outright rejected.
- **Need more information:** These are requests or ideas that you do not yet fully understand, or for which the problem to solve is unclear. After you follow up with the stakeholder who submitted it, you can decide which of the other triage categories it fits into.

---

* One way to help you sort is to use the classic "Eisenhower Matrix," which puts urgency and importance into a 2×2 grid and gives recommended actions for items that land in each of the four quadrants.

Remember that you can only have a limited number of "high priority" items, so you can't put everything into the top bucket. But at the same time, these categories only commit you to "considering" the request, they don't commit you to execute on them right away. In Step 4 (decide on a plan) you will determine the actual next steps for the items you've just triaged.

As part of the triage process, you can also look for patterns. For example, maybe there are multiple ideas that relate to the same problem or initiative or customer, and you can group them together. Maybe many ideas come from the same individual stakeholder or team. These patterns will be useful as you evaluate how to handle all the ideas, and how to iterate on the process itself.

Step 3: Follow Up

For the items you've triaged into the "need more information" bucket, you should work directly with the stakeholder who submitted each request to collaboratively decide which of the other triage buckets it belongs in. It's important to note that while you should involve your stakeholders as much as possible in this process, the ultimate decision of which items go on the roadmap (and when) should belong to the product manager. Follow-up conversations should center around two main concepts: prioritization and trade-offs.

Prioritization is the process of determining the relative order in which initiatives should be completed. Discuss with your stakeholders various factors about each request, such as value, impact, importance, and urgency.* Work with your stakeholders to rank their requests against what is already on the roadmap.

Trade-offs are a logical partner to prioritization. When you add a new item to the roadmap—assuming your resources are not changing—something else has to be removed or delayed to make the new item fit. Having that frank conversation with each stakeholder is important, because, although it seems simple, stakeholders don't always realize that trade-offs must be made.

* Some people like to prioritize using tools like RICE (Reach, Impact, Confidence, Effort).

Use the "ADVISE" framework to help you remember the key components for prioritization and trade-off conversations with your stakeholders (Figure 6-2). ADVISE stands for "Allocation, Deferral, Value, Implementation, Sequence, and Emergency."

Figure 6-2. Use the ADVISE framework to help with prioritization and trade-offs

	Main idea	Examples
Allocation	We have limited resources and we have to be selective about what we work on.	"I wish we could do everything, but our resources are static." "If we could add more resources, we could do both these things, but currently we can only do one of them."
Deferral	If we add something, we have to remove something else.	"What do you think would make sense to push to next quarter?" "Which of these do you think we should do now, and which can wait until later?"
Value	We should prioritize the highest-value items based on how much they move the needle on our objectives.	"How would solving this problem affect our objectives?" "How would you quantify the value of this idea?"
Implementation	Our initiatives must be technically feasible and we should gather the required information before we start.	"Is this technically feasible?" "Do we have all the information we need to design the solution?"
Sequence	We should prioritize our initiatives in a logical order, considering impact, dependencies, and efficient use of resources.	"How would you rank this relative to the other items that are already on the roadmap?" "Are some of these required before others can begin? Do some make others easier?"
Emergency	Some things simply cannot wait, e.g., security, safety, or compliance issues.	"What is the effect if we wait until next month or even next year?" "What is the effect you expect and the timing of that effect?"

Step 4: Decide on a plan

Once you have triaged the requests, it's time to decide on a plan for each one. It's best to communicate your decision clearly to stakeholders, because a "no" now is much better than a long string of "maybes." When you have to say "no" to a request, make sure your stakeholder understands that it's not about "good ideas" versus "bad ideas." You can even mention some key merits of the idea, like "I like how this idea ties together two of our key objectives" but then clearly say why it's not being considered right now. If you've properly discussed the prioritization and trade-offs with your stakeholder, the decision here should not be a surprise. If you get pushback you feel you can't handle, you should escalate to your manager.

You're unlikely to have a fully defined plan immediately, so you'll need to keep your stakeholders updated on the status of each idea in your tracking tool of choice. Here are some examples of the status of an idea:

- **Submitted:** The idea has not yet been reviewed.
- **Following up:** The idea requires additional information and someone has been assigned to investigate.
- **Scheduled:** The idea has been assigned a specific place on the roadmap (you may want to link to a specific theme, epic, user story, or ticket, if you have one).
- **Will consider soon:** This idea has been deferred to a future roadmap planning cycle.
- **Will not consider:** This idea will not be worked on.
- **Duplicate:** The idea has already been submitted, or it is similar enough to another idea that they have been combined.

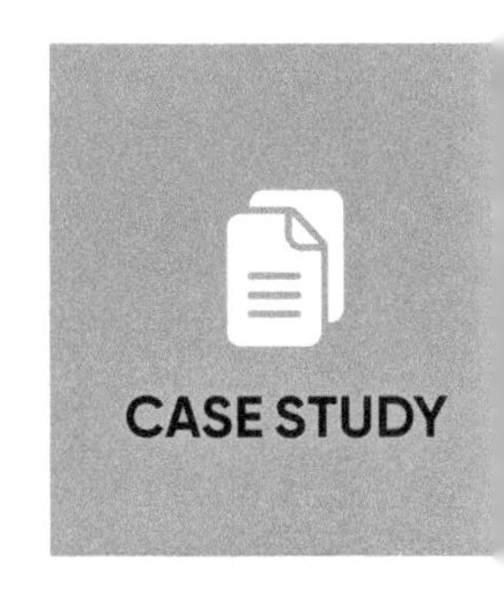

Sometimes an idea is simply impossible. When Melissa was a mechanical engineer, a client asked her to build a machine to collect energy from gym-goers riding stationary bikes. When Melissa reported the maximum energy a bike could produce, the client asked her to double it, which would have produced more energy than was put into it. She asked how they recommended evading the laws of physics and they said "Gears!"

PRO TIPS

What Constitutes an "Emergency?"

Sometimes a request solves a problem that's urgent enough to disrupt the current roadmap cycle, even work in progress. While this should be rare, these cases warrant immediate review with all relevant parties, including your Product Team and the stakeholder who made the request. Use these conversations to determine whether it's actually a "drop-everything emergency" or whether it can wait until the next planning cycle. Here are some questions you should ask to decide if you should drop everything and work on this new request:

- **Cost of delay:** How long can this wait until we need to work on it and what is the impact in the meantime?
- **Opportunity cost:** What else are we working on right now? What are the risks of stopping the current work? How much time would it take to have the team switch from what they're working on now to the new request?
- **Risks:** What are the risks involved in delaying the request? Do we have a mitigation plan? If we don't do this right away, what's the worst that could realistically happen? How likely is that?
- **Dependencies:** What else depends on the work currently in progress? Are there follow-on effects of stopping the current work? What business endeavors might depend on this new work being done? Are we ready to work on the new request right away?

Once you have answered those questions, it should become clear how quickly the new request should be worked on, and whether or not it's really an "emergency." If you have frequent emergencies, consider changing your roadmap planning process. You can make it easier to defuse emergencies by proactively engaging with stakeholders to understand what's coming up and account for it, and maintaining alignment on objectives.

Irie Creates an Intake Process

Irie works with Christina, Eitan, and Min to create an intake process. Once they have the basics sketched out, Irie summarizes. "We'll use this form to collect ideas and requests. If anyone comes to you with something that's not already on the roadmap, you ask them to fill out the form. Then once every two weeks, the four of us will meet and figure out how to handle the requests. It's best if we do this together so we can pool our knowledge and give every idea a fair hearing."

"Can we send all the requests to the intake form?" asks Christina. "Like even requests that come from customers?"

"If it's a request for new functionality, then it goes into this intake process, no matter where it comes from. If it's a bug, or a major customer problem with something in production, you should send them to the regular bug-reporting process with the customer service team."

"What if someone doesn't want to fill out the form?" asks Min.

"Someone?" asks Christina with a grin.

"If 'someone' tries to go around the system," says Irie, "you can send him to me. I'll ask Yacob, Divya, and José to do the same."

Irie makes the rounds with key stakeholders, seeking their input on her proposed intake process, iterating and clarifying things before announcing anything. Sergey agrees readily, happy to have a conduit for input from Alex's research efforts. Ella is inspired by this effort, agreeing to meet with her sales and support teams before Irie's triage sessions to flag their most urgent requests. Philippe even helps her refine the prioritization formula she'd proposed, tying everything back to their product objectives and a high-level estimate of effort.

Since everyone's schedules are busy, Irie decides to roll out the new intake process through individual meetings, rather than having a group presentation on it. She manages to find time with Sparks to review the new process. He seems to accept it, but she has her doubts. *I guess we'll see how it goes*, she thinks.

There's one more person she has not yet spoken with—Liz. The CEO hasn't requested anything directly, so she's not sure how relevant the intake process is for her, but maybe Liz can help her enforce the new process with Sparks. She wants to review the new roadmap with her anyway, so she decides to address both in one meeting with Liz. ■

6.2 Irie Updates the Roadmap

Irie asks Pria if she can get some time with Liz to talk about the roadmap and the new intake process. She has spoken with Helthex's CEO a few times, despite Liz's reduced schedule. She's been out for longer than expected with her health issues.

Irie dials in from home and immediately notices Liz looks tired around the eyes. She asks Liz how she is doing.

"Doing better," says Liz. "Sorry about the 6 p.m. meeting time. I'm trying to navigate all the doctor's appointments and still find time for work."

"No worries," says Irie. "It must be hard trying to get back to work with all that going on."

"Yes, it is," says Liz. "But that's part of the reason I'm excited to come back. Our product is meant to help people dealing with complex medical problems."

"That's right," says Irie, agreeing.

"Speaking of which," continues Liz, "you mentioned having a roadmap to show me?"

Irie shows her roadmap, explaining the different themes and how they tie in with the company's objectives.

"This is really great," says Liz. "I love the OKRs, they're really simple. Much better than the ones we had last year."

"Thanks," says Irie. "Some of the details are going to change a bit, actually, but the OKRs and the overall strategy are staying the same."

"What's changing?" asks Liz.

"Well, Sparks has been pushing for a focus on AI," says Irie. "I guess it makes sense, given that our app makes suggestions based on a lot of different pieces of data."

"You don't seem sure of that," observes Liz. "Do you think we shouldn't be thinking about AI?"

"No, we definitely should. It's backed up by Alex and José's research. But Sparks seems to be pushing for more and more AI features for the sake of 'doing AI.' I'm worried that we're going to focus too heavily on that, and possibly undermine our overall product."

"Hmm," says Liz. "Sparks tells me we can do the AI stuff in addition to the other work."

"We can do some of it," says Irie. "But there's an opportunity cost to adding a whole new workstream to an already tight roadmap. We can't work on everything, so we need to consider the trade-offs. Sparks seems to think we can magically do more in the same amount of time, with the same resources."

Irie pauses. She is worried that she may be coming off too strong with Liz, but she's trying to balance that with getting a lot of information across in the small amount of time she has with her.

Liz says, "I hear that Sparks has been a bit difficult lately."

"I guess you could say that," says Irie, not sure where Liz is going with this.

"He can be a bit pushy sometimes," Liz continues. "But I agree with you both that building smarts into our product is important. We'll figure out a way to get it done."

"Any thoughts on resourcing?" asks Irie.

"I'm not sure," Liz says, sounding uncertain. "Sri said he might have a good agency to use to get us some engineering teams who could ramp up quickly. Divya says she knows someone in India. I've asked Sparks to coordinate that."

Irie suddenly suspects that either Liz doesn't know about Sparks's partnership idea, or she doesn't want Irie to know about it. She decides to dig in a little.

"I'll let you know if I have other ideas," says Irie. "Did Sparks mention anything about the integrations with LifeWalk, or Alvex? We had a discussion about it today, and he seems really insistent."

"Why the focus on integrations?" asks Liz. "I thought we were doing AI, not integrations." This solidifies Irie's belief that Liz doesn't know about Sparks's partnership plans.

"We need both," says Irie. "Alvex has an AI chatbot and LifeWalk has data we can use to train the AI."

"Okay, that makes sense," says Liz.

As they continue their discussion, Irie realizes that AI isn't represented on the roadmap as an overall theme. And, although the OKRs are still relevant, the work they're doing to achieve those goals needs to shift to accommodate the changing company strategy. She realizes her roadmap needs an update already after only eight weeks.

Irie shares her plan with Liz. "Since AI is becoming a much bigger part of our product strategy, I'm going to update this roadmap to include it explicitly. I'll work with Sparks to make sure we're on the same page, and I'll invite you to the roadmap update meeting."

"Perfect," says Liz. "This is just the sort of adaptability we need." ■

Roadmap Planning

To adapt to the changing world around you, your roadmap should be a living document, updated over time. A typical cadence for roadmap updates is quarterly, but different cycle lengths work for different situations. The planning process gives you an opportunity to rethink your hypotheses and respond to findings from your discovery initiatives. This is especially helpful if you are still working on product-market fit, or if your industry is prone to rapid change.*

"Roadmap planning" is an iterative process by which you update your roadmap periodically to respond to changes. Having a robust planning process, including proactive stakeholder engagement, may not avoid all mid-cycle changes, but it should reduce the "emergencies." A roadmap planning process has a few key components: 1) collect input, 2) draft the roadmap update, 3) sort out dependencies, and 4) confirm with stakeholders (Figure 6-3).

Each of these steps should have specific dates on the calendar that are shared by all Product Teams and widely communicated with stakeholders. Creating and sharing your planning process timeline is helpful to ensure your stakeholders understand when the best time is for them to provide feedback and requests. Explain to them that it's easiest for them to provide requests during the planning period because that's when the roadmap is most flexible.

When you put together your planning timeline, make sure you leave enough time for all four components. For example, if you have many stakeholders to collect input from, think about how quickly the product managers on the team can reasonably secure meetings with them. Think about when you want to schedule the final confirmation meeting (step 4) and work backward on timing. As a general guideline, we've found that quarterly planning has to start in the middle of the previous quarter.

---

* Or if your executives are prone to rapidly changing their minds.

Figure 6-3. Roadmap planning steps

Step 1: Collect Input

Gather information and changes since the last planning cycle like team velocity, carry over work, technical learnings, customer interviews, stakeholder requests, strategy updates, and external changes.

Working with stakeholders

Proactively connect with stakeholders to discuss strategy, solicit input, and ask if anything has changed on their side.

Step 2: Draft Roadmap Update

Consider the first version of your updated roadmap to be a draft, because there are several additional steps you should take before it is finalized.*

Working with stakeholders

Iteratively review your draft roadmap with each stakeholder to align on the direction and strategy. Now is the time to discuss why items they've proposed are (or are not) included. It's also helpful to hear what questions they have so you can pre-address those in your roadmap update discussion.

Step 3: Sort Out Dependencies

Review your draft roadmap with other Product Teams to identify interdependencies, overlaps, or gaps in the work being proposed. Align with them on the planning cadence and timeline, to avoid roadmap rework and intra-cycle revisions.

Working with stakeholders

It's important to also sort out dependencies with stakeholder teams who will be doing work alongside the Product Teams. For example, the marketing team likely has go-to-market activities and they need to coordinate resourcing and timing with your plans.

Step 4: Confirm with Stakeholders

After you have aligned your roadmap update draft with stakeholders and other Product Teams, bring all your stakeholders together for a final confirmation of the plan.

Working with stakeholders

The roadmap update meeting should have no surprises for your stakeholders, because you will have previewed† the updated roadmap with them individually. Likewise, the meeting should have no surprises for you, because you will have solicited input and feedback from each of your stakeholders in advance.

* Your roadmap is never really "finalized"—it's a living document that is updated on a regular basis. But for the sake of simplicity and stakeholder expectation setting, it's useful to think of your roadmap as "finalized" at a certain point in the planning cycle so engineering can plan their work with a lower risk of continual change.

† Previewing, also sometimes called "shuttle diplomacy" is when you review a big change or decision with each stakeholder individually before bringing them together as a group. This aids alignment because you can hear feedback and concerns from each person and dig deeper in one-on-one conversations.

PRO TIPS

Roadmap Cycle Length

Different companies use different cadences for roadmapping cycles. Mostly gone are the days of the annual "waterfall" cycle, but in some cases it still makes sense to plan once a year, like with physical products. Here are some pros and cons of each roadmap planning frequency (Figure 6-4).

Figure 6-4. Roadmap planning frequency options

Planning frequency	When to consider	Benefits	Challenges
Annual	Mature and stable market, manufactured physical products, seasonal items, dependency on multi-quarter research and development	Consistent direction and strategy, follow-through on long-term vision, easier to allocate resources	Little flexibility to adapt to changes, potential misalignment with emerging trends
Biannual	Industries with a moderate pace of change	Balance between long-term vision and ability to shift priorities	Some blocks to adapting to change, approach can feel less agile
Quarterly	Dynamic markets with some stability but also regular changes	Flexibility to adapt to customer feedback and market shifts	May incentivize focus on shorter-term goals rather than long-term vision
Monthly	Fast-paced industries, rapidly changing customer needs, startups still looking for product–market fit	Enables quick adaptation to short-term market changes and real-time data and feedback	Resource-intensive due to frequent roadmap planning activities, can lose sight of broader strategic initiatives

Irie Pre-aligns with Sparks

Once Irie is set with her draft roadmap update, she makes the rounds with her key stakeholders to review it. She dreads her pre-alignment meeting with Sparks. At the end of every meeting she feels like she's figured out a way to work with him, but then it becomes clear by the next meeting that Sparks is not interested in changing his views on anything.

"So why are we here?" Sparks says by way of greeting.

Irie cringes a little at this, but takes a breath and answers the question. "I know you don't think we need so many meetings," Irie says, interpreting Sparks's comment, "but I want to make sure that you're on board with the updates we've made to the roadmap since the last time we presented it. I don't want there to be any surprises, so I want to make sure you have the space to voice any concerns before we meet as a group tomorrow."

Sparks agrees and Irie shows him the updated roadmap that focuses more explicitly on AI.

"This all makes sense," says Sparks. "But where are the specific AI features from the presentation I did last week? This isn't granular enough."

"Sparks, we've gone over this," begins Irie, trying to stay calm. "We can't do everything immediately."

"Right, but if we had a partner with resources and part of the technology, we could do more, right?" asks Sparks.

Irie's ears prick up. She thinks he might be willing to reveal more about his partnership plan. She wants to see if he'll open up, so instead of answering his question, she asks him questions instead.

"Like what kind of partnership?" asks Irie.

"I think Alvex's chatbot can take most of the load here," Sparks says.

"With any partnership, there's something to build," says Irie. "There's no such thing as a pure 'off the shelf' solution. There's always some integration work, at least."

"Are you telling me we can't do any of it at all?" says Sparks, raising his voice. "What's the point of saying we're focused on AI if we have nothing specific to back it up? Do I have to tell the board that we can't deliver what we promised?"

"No," Irie clarifies. "I am saying I can't give you features and dates yet with no idea how we're going to deliver."

"Just trust me that we'll have what we need," says Sparks.

"You're asking me for a best-case roadmap," begins Irie. "But that's where broken promises come from. Every time in my career I've just assumed things would work out I've been burned. That's taught me to do my homework before promising."

The pair are silent for a moment, seemingly at an impasse.

Irie says, "If you could give me details about what Alvex or whoever is proposing for support, that would give me more to go on."

"I can't go into details right now," Sparks replies. "It's delicate."

Irie feels she needs to buy time to figure out how to respond. "It looks like we're at the end of our meeting time," she says. "We can discuss this more another time. By the way, Liz asked if she could join the roadmap review, so I added her. She wants to start to get back into things and we thought that would be a good opportunity."

Sparks doesn't answer immediately. He is visibly confused and possibly annoyed. "Okay, fine," he says finally. "See you tomorrow."

The next morning, Irie begins the meeting cheerfully. "Welcome, everyone, to our first quarterly roadmap update meeting. We're going to be doing these every quarter, to go over changes to the roadmap and make sure we're all on the same page."

Schedules were tight, so they are meeting on a Friday, when most people work from home. Assembled in the virtual meeting are the technology team leaders (Sri, Irie, Yacob, José, and Divya), Sergey, Arianna, Ella, Philippe, and Liz. Sparks is dialed in too, but he is sitting by himself in a large conference room in the office. The video is zoomed out so it is hard to see his facial expressions.

Irie is relieved that Liz is able to attend. She is hoping the CEO will have a moderating effect on Sparks.

"I've met with all of you in advance of this meeting, so I don't expect any surprises, but we're here to have any discussions that are necessary."

Irie goes through the presentation, reviewing the OKRs, the initial plan, and the updates. She meets no objections.

"Sparks," says Irie. "I want to make sure you're on board with the updated roadmap. You expressed some concerns when we spoke."

Sparks looks up from his laptop when he hears his name. "No problems," he says. "I do have a concern about resourcing, though. There is a lot more we need to add to the roadmap."

Sri speaks up for the first time in the meeting. "We talked about outsourcing either here or in India. I have asked Yacob and Divya for proposals."

"I don't think that'll be necessary," says Sparks, "If either of these partnerships with Alvex or VigGuard come through, we'll have all the resources we need."

Irie looks at Liz's image on her laptop. She seems unconcerned, even unsurprised. "Liz," Irie asks, "do you have any concerns here?"

"I really like the direction things are going with adding AI functionality to the product. Irie, you've really shown that you can get the team into shape, and you present the roadmap in a well-organized way. Maybe you can work closer with Sparks on these partnerships? He could use the help figuring out how to get all the things we need onto the roadmap."

Irie's not sure what to make of that request. Should she be honored that Liz thinks highly of her work? Or should she be worried that Liz basically just gave Sparks permission to put more stuff on the roadmap? ■

6.3 Irie Creates a Roadmap Routine

Irie is meeting with her product council every two weeks for thirty minutes because so much is changing as they try to incorporate the new AI strategy. These meetings are helpful to bring up any concerns and feedback the team has, and discuss any new requests. Sri is out with a sprained wrist and is at a doctor's appointment during today's meeting.

This week Sparks gets the meeting started. "I have a set of new things to add to the roadmap, so I'm going to share those with you and I need estimates this week."

"Sparks," interrupts Yacob, exasperated. "You know we have a process for new ideas like these."

"These aren't ideas," says Sparks. "They are requirements. They came directly from Alvex and VigGuard."

Looking around the room at unhappy faces, Sparks adds, "You can put them in whatever system or process you need when we're done."

"Okay, let's hear what they are asking for," says Irie, trying to be accommodating. "But I can't guarantee anything."

Sparks shares his screen on the display in the conference room and brings up a list of 17 items.

"That's a lot," says Irie. "How about sharing your top three?"

"This is already the top list! My partners had almost 50 items. These are just the critical ones."

Irie, Yacob, and Divya exchange glances as Sparks goes through the list. Yacob sits back with his arms crossed, like he's trying to protect himself from the onslaught of new assignments. Divya mutitasks on her computer, pretending to take notes. Irie stares at the screen, wondering how to get Sparks to start following any reasonable sort of process.

The group lets Sparks keep talking until he gets to Number 7. "This is a good one: Integration with JogWave. This will let people connect their health routines with our AI engine, so we can better recognize patterns. VigGuard provides JogWave for free to their employees. A lot of them are using it, so we could use that vast amount of data to train our AI. Okay, Number 8—"

"Wait," interrupts Divya. "I thought we were using SmartStride data to train the AI for exercise patterns?"

"We talked about that," says Yacob, "but JogWave has a bigger customer base and more types of exercises, so we thought that would be a better choice. We're already working on it."

"But my team has been looking at SmartStride data," says Divya. "The version of the roadmap that I have says SmartStride."

"We changed it a few weeks ago," says Irie. "I didn't realize your team had already started working on SmartStride. The last time we spoke they were still setting up the new modeling environment."

"Right, but we finished that and moved on," says Divya.

"The good news," says Sparks, "is that you're all on board with the JogWave integration, and we can get it soon. We just reduced my list to 16!"

"I'm not on board yet," says Divya.

"Okay, I think we need to do some roadmap check-ins between planning cycles," says Irie. "Clearly, I didn't do a good job in communicating. Sparks didn't know we were already doing this integration, and your two teams aren't even working on the same things."

Yacob offers a suggestion. "Why not just have a monthly roadmap review meeting? We used to do those at my last company. They were very helpful to make sure we stay aligned over time."

"I did those at my last company too," Irie says. "I'll set that up. We should invite other people who are not on the product council to that one. I'll put together a list."

"For now, though, where do we stand?" asks Divya. "Did we just waste a bunch of work?"

"I don't think so," says Irie. "We were planning on doing multiple integrations anyway, with SmartStride and JogWave being the first two. We should just pick one to go first and stick with that. It sounds like Sparks is a bigger fan of JogWave. I'm open to that. What do the rest of you see as reasons to go with one or the other?" Irie is trying to throw Sparks a bone so it seems like at least they're doing one of his ideas.

"I think we need to see what the options are before we decide. I'm going to do some digging and see what it would take to switch to JogWave," says Divya. "It will definitely put us behind, but there's no reason we couldn't switch, to line up with what Yacob's team is working on. In the future, it would be really nice to let me know, though."

"Of course," says Irie. "That was my fault. The roadmap review meeting should help keep everyone aligned between quarterly planning cycles."

"That would be really helpful," says Divya.

"I'll be there too," says Sparks. "I clearly can't trust you all to make good decisions."

Sparks continues with his presentation. "Okay, so Requirement Number 8…"

Irie stares out the window while Sparks speaks. She's starting to disengage and she doesn't like how it feels. ■

Routine Roadmap Reviews

While the roadmap planning cycle gives you the opportunity to reconsider the big questions, within the roadmap cycle it's best to preserve the plan as much as possible so as not to negatively affect work in progress. Context switching slows down delivery, and making major updates in the middle of a cycle is very disruptive. But how do you square that with the "continuous discovery" that you're supposed to be doing? What about "test and iterate"? What if something changes? While major changes are disruptive, minor changes are normal and should be expected and planned for.

Responding to changes

There are many different types of changes in the world that might drive you to update your roadmap between planning cycles. Changes come in all different shapes and sizes (Figure 6-5). Major changes sometimes call for complete roadmap shakeups, but minor changes generally don't alter the overall objectives, so they can usually be handled through a simple update to the existing roadmap.

For major changes, you may need to recreate the roadmap from scratch and get alignment on a whole new set of priorities. But those major changes don't happen every day or even every quarter. Because minor changes are bound to happen frequently, you should build an update process into your routine to handle them in a predictable way.

Figure 6-5. Types of changes that might cause you to update the roadmap

Change in the world	Minor example	Major example
Competition	A competitor comes out with a surprise feature, but you're not sure if customers will want it.	A new competitor enters the market and your customers immediately start migrating to their product.
Resourcing	One engineer resigns from a team of 25.	A major investor pulls out their funding.
Learning	Customers use one feature in an unexpected way.	The current architectural plan will not meet the security standards of the industry.
Executive	An executive just learned about a feature you're working on and has some great suggestions.	A new CEO has joined the company and wants to rethink the strategy.
Market Factors	A new trend has slightly altered customer behavior.	A new technology enters the market and makes your technology obsolete.

PRO TIPS

Alignment Decay

Regular communication with stakeholders is important when you have to update your roadmap, but it's equally important even if nothing changes. People naturally tend to forget about the roadmap over time: they forget what the plan was, they forget what they agreed to, and sometimes they even forget what the goals were. Over time, stakeholders can gradually move apart in their opinions. This can lead to "Alignment Decay," where alignment among stakeholders fades over time.

Alignment Decay happens when people initially align on a plan, and then go back into their silos to work on it, without communicating along the way. If decisions are made independently without consulting other affected people or teams, plans can get quickly out of sync. This risks creating products that simply don't work, don't solve the customer's problem, or aren't marketed and sold properly.

The key to preventing Alignment Decay is consistent communication. Remind people what you're working on, what they agreed to, and why. If your stakeholders haven't heard from you in a while, they may go so far as to assume there is no roadmap anymore...and they may create their own plans. Frequent communication, such as sending out regular update emails, can help here. To remind people of the plan, start each email with one or two sentences that reiterate the goals, the plan, and the agreements.

And, of course, when things do change, communicating with the entire team is essential. For example, if something changes in the market, and the sales team updates their sales pitch but doesn't tell the product management team, then they could be promising customers features that are not in the plan. To prevent this, make updates to your roadmap readily available and give your stakeholders ample opportunities to provide feedback and ask questions. One way to do this is by establishing roadmap review meetings.

Roadmap Review Meetings

Regular roadmap meetings with stakeholders are critical to maintain alignment over time and prevent Alignment Decay. It also forces you to periodically get out of your day-to-day work and confirm stakeholder alignment. Roadmap review meetings have a few key components:

Invitees

Invite your Product Team, key cross-functional stakeholders (e.g., your product council), and Power Players. Use your Stakeholder Canvas to ensure good coverage. In large companies, you may need ten to fifteen stakeholders in the meeting; in small companies, you may only need your product council. In an organization of thousands, or for specialized products, you may need more functions, like professional services, customer support, or legal.

Cadence

Hold roadmap review meetings periodically during your planning cycle. For example, if you plan quarterly, you should review monthly. Setting up recurring meetings in advance and sending out reminders will help with participation. Make sure your key stakeholders are willing to attend the series or to send a delegate.

Agenda

Set expectations by sending out the agenda in advance of the meeting. A pre-read is especially helpful when there are many attendees or when there are important changes to discuss. A good roadmap review meeting agenda includes the following:

- **Review the goals.** Ensure alignment on the objectives before discussing updates.
- **Review the roadmap.** Give a brief overview of what you've been working on and why.
- **Talk about successes.** Share your wins, and what you've learned since the last review.
- **Discuss updates.** Review changes, their drivers, and the impact. Confirm alignment on the path forward.

Pre-alignment meetings

An effective roadmap review meeting should be simply a stamp of approval. Meet in advance with each stakeholder individually to learn about their needs, make adjustments, and confirm alignment (also called "shuttle diplomacy").* This can also prevent a "pocket veto"—where someone seems to agree in a group session but then acts differently on their own.

---

* Bruce's book, *Product Roadmaps Relaunched* (O'Reilly, 2017) describes shuttle diplomacy as "meeting with each party individually to reach decisions that require compromise and trade-offs."

Decision Logs

Another way to maintain alignment is to write down what was decided and when—usually called a "decision log." If it isn't written down, people may forget what was decided (deliberately or not). Here are some example decision logs formats and when to use them.*

Detailed decision log

When massive, important decisions are made, it can be helpful to capture the relevant details about the decision. Use these for decisions such as a major pivot in the product, funding a brand new team, or deciding not to do a feature requested by a major customer.

In these complex situations, it helps to document as much information as possible, so that if you need to come back to it later, you will remember exactly what happened. Components of this decision log often include the following:

- Date—date the decision was made
- Names of the Driver, Approver, Contributors, and Informed people†
- Problem statement—why the decision was needed
- Assumptions—what informed the decision
- Decision made—what was the actual decision
- Reasons—why this was the decision
- Desired outcome—expected benefits of the decision
- Risks and mitigations—what could go wrong, and how we hope to prevent it or minimize the effects
- Check-in date—when you plan to decide if the decision has achieved its desired outcome
- Assessment—data to be assessed on the check-in date

---

* Visit *alignedthebook.com* for more information on decision logs.

† See Chapter 1, Section 4, "Irie Determines Decision-Making Authority," for more details on the DACI model.

A key benefit of a detailed decision log is that the check-in and assessment sections encourage you to come back later and evaluate whether the decision was a good one, which can help you make better decisions in the future.

Simple spreadsheet

For everyday decisions, create a spreadsheet that lists a few key elements of each decision:

- Questions that were asked and their answers
- Details of the path forward
- The date the decision was made
- The name of the decision maker

Even something this simple can serve as a helpful resource for the entire team when they are trying to remember what was decided and when. Although it captures less information than the detailed decision log, having fewer required pieces of information can increase compliance.

RAID log

When you want to log not only decisions, but also other aspects of the initiative, you can use a RAID log. RAID stands for "Risks, Actions (or Assumptions), Issues, Decisions (or Dependencies)." Writing down not only decisions that are made, but also risks that are identified (and mitigation plans), actions that need to be taken (and by whom and by when), and issues (that need to be discussed) can be a helpful way to keep track of the many aspects of a complex project.

Decision messages

In addition to the log entry, it is helpful to summarize decisions in writing, like an email or Slack message, and send them to key stakeholders just after the decision has been made. With a decision message, you can get immediate feedback on whether or not the decision details were captured correctly. For example, after a meeting in which one or more decisions were aligned on, you can send out an email to the meeting attendees that recaps the decisions and asks if anyone has any corrections. This will also serve as evidence that the decisions were in fact agreed on at the time.

Irie Reviews the Roadmap

Irie decides to set up time with each stakeholder before the first monthly roadmap review meeting, just like she did with the quarterly roadmap update meeting.

To prepare for the meeting with Divya, Irie puts together a pros and cons list for the decision to switch Divya's team from SmartStride to JogWave. She doesn't feel that either integration is inherently better, but she wants to demonstrate what a good process looks like. And maybe giving Sparks a win will make him more cooperative, so she will try to convince Divya to change what her team is working on to match Yacob's team.

Irie wants to make small talk as they join the video call, but time is short, so she cuts to the chase. "Listen, about the misunderstanding on the roadmap, I really want to apologize. There's so much going on that I guess I forgot to tell you about the change in the integration partner."

"Actually," says Divya, "I need to apologize, too. My team *was* working on the JogWave data, not the other one; they just didn't tell me they'd switched. Your team and my team are on the ball, even if we're not. I guess that's what happens when you're a bit removed from the day-to-day life of the team."

"Well that's a relief," says Irie. "At least we're aligned there." She puts away the pros and cons list she prepared. "I also wanted to talk about resourcing. I keep getting pressure to speed things up, but we can't do everything with the team we have. I met with Liz yesterday, and she mentioned that Sri has an agency in the US—and you and I discussed ideas about staffing in India."

"Yes. Hiring in India just makes sense. It's more cost effective, tech workers in India are incredibly talented, and it would be internal employees, which would help with motivation."

"Internal employees?" asks Irie.

"Yes," says Divya. "I'm envisioning opening up a new office in India, not just contracting. It would make the most sense."

"Why open a new office, though?" asks Irie. "That seems like a lot of work when we could contract. We're not even sure how well these AI features are going to go. I mean, initial feedback is good, but it's still a bit of a risk to invest in an office right now, no?"

"I don't see it that way," says Divya. "My brother is in Pune, for example, and he's a capable GM who could open an office quickly."

Irie suddenly understands why Divya is so intent on opening an office in India. She wants to get her brother a big job! She doesn't want to embarrass Divya or make her defensive, though, so she makes a mental note of this new information and moves on.

"Okay," says Irie. "Well, however we staff it, you do agree we can't do everything Sparks wants with the team we have, right? I'm hoping I can get your support on adding to our team if we want to do this work, not just hoping we can miraculously fit it in."

"Totally agree," says Divya.

Irie's next preparation meeting is with Sri, Yacob, and José. She needs their help to refine the narrative for the roadmap review meeting.

Irie begins by asking about Sri's wrist.

"It's getting better," says Sri. "That'll teach me to go with my gut instead of letting my nephew convince me to go rollerblading. Sorry I've been in and out of the office with these injuries lately."

"It's okay," says Yacob. "We got you."

"I know you do," says Sri. "It's great to have a team I can trust."

"I appreciate that. But before we get all mushy," says Irie, smiling, "I do have actual agenda items."

"Of course," says Sri, with his trademark smile.

"We have our first roadmap review meeting tomorrow," Irie begins, "and I want to make sure we're all aligned on what we're doing. Yacob and I met with Divya and Sparks the other day and it seemed like we weren't aligned on the basics, like whether we're starting with the JogWave or SmartStride integration first. Sparks didn't know we were doing an integration at all, and Divya thought we were doing a different one."

"I'm sorry I wasn't there," says Sri.

"No problem," says Irie. "I spoke to Divya, and it appears that we *are* actually aligned; she was just out of the loop. Which is a relief, but it points to the need for better communication. So I want to take another step back and make sure this group is aligned internally before we talk with stakeholders."

"Thank you," says José. "I feel like design is left out sometimes."

"Not anymore," says Irie. "Product, design, and engineering are a team. And we all need to be aligned on the roadmap before we present tomorrow. I have a draft of the presentation here, and I'd love your feedback. This month our updates all seem to be trade-off decisions, so I framed the slide that way."

Irie displays a slide with updates since the quarter started.

Figure 6-6. Irie's roadmap update slide

HELTHEX

Roadmap Update

We have postponed...	And are focusing instead on...	So that we can...	Risks/notes
User Access Management	JogWave integration	Have a data source for AI modeling for exercise behavioral indicators	This change will further delay a required feature for large customers
Upgrading platform to DataTap 4.5	Stability enhancements	Reduce downtime and latency in the app	We discovered this problem in testing after the roadmap was set for the quarter; DataTap 3.0 support ends in 6 months
Starting new features for the Admin Console	Fixing customer issues with user account management	Have UAM ready for RightBank contract renewal	We decided to get this feature right before moving on to other features

"Irie, this is great," says Sri. "It's easy to understand what we changed and why."

"Thank you," says Irie. "Anyone see anything we're missing?"

"We also decided to add the discovery work for SmartStride, correct?" asks Yacob. "I'm having some of my folks look into that, just to see how similar it is to JogWave. Timeboxed, of course."

"Good, I'll add that," says Irie. "Have we taken something off the discovery list in order to add that in?"

Yacob thinks for a moment. "I suppose we did. We're pushing off some technical work because of a dependency on upgrading the platform."

"Okay, that makes sense," says Irie. "Have you put those updates into the new decision log?"

"Yes," says Yacob. "It's really helpful. The Product Teams add to it in meetings like standups and planning sessions, and then it's available for anyone to look at. We've already found one place where a decision was made that affected another pod and they were able to find that problem early and fix it."

There are no more issues to discuss, so they wrap up the meeting. And with all the preparation Irie has done, the roadmap review meeting goes off without a hitch. ■

6.4 Irie Receives More Requests

Irie gets a chat message from Sparks.

Sparks: I need you to add a facial recognition feature.

Irie: That's interesting, tell me more.

Sparks: We need it for the JogWave integration

Irie: What's the problem we're trying to solve?

Sparks: Just add it, please.

Irie: I need to at least know more about what you want. "Facial recognition" is a broad request.

Sparks: OK, I'll just go to engineering.

Sparks: [Five minutes later] They sent me back to you. I would get Sri to put it in the roadmap for me, but he's out today. So I guess I'm stuck asking you.

Irie: Why do you need facial recognition, what are people going to use it for?

Sparks: Signing in.

Irie: OK, that makes sense. You just want us to integrate Apple Face ID with the app login? We had planned on that, but we decided to fix a high-priority customer bug instead. It's on the list for next sprint.

Sparks: OK good.

Irie is happy that her team sent Sparks back to her. If he's not going to follow the intake process, at least he can stop bothering the whole team.

Sparks then asks to add a feature for users to share their results with friends. Irie suggests they discuss, so they set up a quick video call. Skipping the small talk, Irie asks how this idea matches with the product objectives they aligned on: retention, issue reduction, paid conversion, and use case coverage.

"I guess I can see it being paid conversion," says Sparks.

"What evidence do we have that sharing with friends will improve paid conversion?" asks Irie. "Are you proposing we charge users for it?"

"Why do these things matter so much to you?" asks Sparks angrily. "What do I have to do to convince you to just do them?"

Irie takes a deep breath before answering as calmly as she can. "I'm not asking these questions to be difficult. I'm asking because I care about the success of the product, just like you do. And I know that the best chance we have for success is to focus on what will drive results. If we don't do that, we'll be spinning our wheels working on things, but never delivering anything of value. We need ROI on our efforts to create a valuable and successful product."

"ROI…" begins Sparks. "Okay, I see your point. But we need a long list of things to make the Alvex and VigGuard partnerships work."

"Partnerships are not part of our objectives," Irie points out.

"Some things go beyond the normal objectives and priorities," Sparks says.

"If it's that strategic," asks Irie, "shouldn't Liz know all about it? When I talked to her last week, she didn't seem up to speed."

"I don't want you to bother her with this stuff," says Sparks. "She has enough to worry about. And you really shouldn't have invited her to the review meeting the other day," he adds.

"She said she wanted to come when I spoke to her," Irie says, reasonably.

"She takes on more than she should," Sparks replies. "Please don't tempt her and leave these things to me."

After they hang up, Irie calls Sri, who is still out with his wrist issues.

"I have never had this much trouble with a stakeholder before," Irie shares. "I just can't figure him out." She recaps the conversation for Sri, including keeping the details away from Liz. Then she adds, "He acts like I'm some robot he can control!"

"Sometimes we just need to say 'no' to people like Sparks," says Sri. "But he's never been easy to say no to."

"Have you ever found a way?" Irie asks.

"I mostly give him choices," Sri replies. "We can do this or that, but not both in a given time with the resources we have."

"I've tried that," says Irie. "He keeps asking 'but what if resources were available,' and I'm not sure how to respond to that." After a moment, she asks, "Didn't you tell me once that you talked him out of expanding into Canada?"

"Yes," says Sri, "but it wasn't just me. Sergey figured out the market was one-tenth of the US, Alex uncovered a lot of different regulations we'd have to comply with, Christina had a whole list of additional data integrations we'd need, and, given all that, Philippe's business model showed it was a loser. It took the whole team to convince him."

"So it was the analysis that did it in the end," Irie says. "The cold, hard facts."

"I dunno," says Sri, frowning as he thinks. "Sparks has a way of ignoring inconvenient facts."

Irie rolls her eyes at this but is forced to agree. Sri continues. "I think it was the weight of everyone uniting in their opposition that did it, really."

"So I'd have to get everyone to line up against these things," Irie says. "These things that I'm not even supposed to talk about."

"Or maybe you could go to Liz," Sri says.

"But I'm not supposed to talk to her either," Irie says, feeling helpless. "Perfect."

After her conversation with Sri, Irie decides to call Darius to see if he has any advice on her no-win scenario. ■

How to Say No Tactfully

Ideally, when you need to say "no," it's because you and your stakeholders have collectively concluded that "no" is the best course of action. But sometimes you need to be the bearer of bad news. You may need to say "no" because implementing the idea could be harmful in some way, or it could just be a good idea that's very low priority.

If you have a particularly difficult stakeholder, you may want to craft a persuasive narrative before you speak with them. Thinking it through in advance will help you communicate the reason behind the "no" in a more compelling way. Think about their decision-making style and their departmental incentives to structure your "no." What types of arguments would make them eager to accept the "no" rather than fighting back?

A product manager we know was working on a data quality project and was having trouble explaining the benefits to sales and marketing executives. Their reasoning was that customers weren't asking for data quality, so why should the team spend time improving it? Understanding that sales and marketing stakeholders care about conversion rates and customer retention, the product manager changed his narrative. Instead of "data quality," he started talking about "customer confidence." This got the sales and marketing executives on board with making technical improvements to the product's data platform, instead of working on their new feature requests, because it now aligned with their goals.

If you're uncomfortable saying "no," you should work out the details ahead of time and practice what you will say. Create a narrative that shows that "no" is the best (or only) choice, and make sure your story is compelling. By now, you should understand enough about your stakeholders and their concerns to know why a particular request is so important, so that you can let them down easy. Figure 6-7 offers some tips for when you have to say "no" to a stakeholder.

Figure 6-7. Tips on saying "no"

Reframe

If you phrase it as "this or that" instead of "no," it's easier to make a well-considered choice.

Example

"We could add that feature now and skip the security work, but we risk a lawsuit. If we delay the release to do both, it's a much safer situation. I recommend the second option."

Be transparent

Build trust by providing clear reasoning and walking people through your process. Better yet, involve them in that process.

Example

"I came up with three different options for fitting this feature into the roadmap, but there are trade-offs. I'd like to walk you through it and maybe you'll see something I've overlooked."

Bring emotion into the decision

Adding evidence, like customer quotations, can help your stakeholder empathize.

Example

"If we do this new feature, we won't be able to continue to improve the existing features. One customer said, 'I'm spending half my day clicking around. It would be so much better if I had a bulk upload option.'"

Take emotion out of the decision

If you bring data into the argument, you can more easily make a decision without a battle of opinions.

Example

"87% of our customers use this feature daily, so we really should test the changes with customers before we release them."

Bring emotion into the conversation

Show empathy, transparency, and understanding for your stakeholder's situation.

Example

"I don't want to have to say 'no' either, because I know how important this is to you, to me, to everyone. But we both know it's the right choice for the situation we're in."

Take emotion out of the conversation

Focusing on what the right decision is, instead of simply what people want, makes it clear that this wasn't a personal choice.

Example

"I know this is frustrating for our sales people, but unfortunately we have to choose the option that generates the most revenue, not the one that saves people time."

Bring an SME for backup

You can ask a subject matter expert to help you explain the situation.

Example

"I've brought John with me, just in case you have any more technical questions, because he's the engineer who built the system."

Confirm the decision

Ask your stakeholder to confirm the "no" decision, and be open to considering a different opinion.

Example

"I've laid out my reasoning, but I'm happy to be wrong. What do you think the right decision is?"

Save your opinion

Make the decision without bias.

Example

"I have laid out the impact of adding this feature to the roadmap versus not adding it. I have an opinion on what to do, but I'd like to hear from you first."

Suggest an alternative

Your stakeholder may actually be happy, or even happier, with an alternate solution.

Example

"I can't do that right now, because of dependencies. But what I can offer you is X, and here's why I think it's even better..."

Simplify technical issues

When the reason for saying "no" is too technical for the stakeholder, explain it in simpler terms.

Example

"The platform team has run into problems where they have to rebuild part of the platform to send this particular type of message, and that puts the new messaging feature back by at least another month."

Register the request for later

If you can't do the request now, but you still think it's a good idea, make sure they know you heard them by putting the request somewhere visible.

Example

"I know we both want to be able to do this right now, but we've agreed it's less important than the work we're currently doing. I'm going to add it to the top of the list for consideration for next quarter."

When to Escalate

If you are having trouble saying "no" or your stakeholder isn't accepting "no" for an answer, you should escalate. It is critical to say "no" as soon as you know that's the answer, because setting expectations early is important. It's better to say "no" and disappoint them *now*, than to say "yes" and not be able to deliver, disappointing them *later*. So when you realize your "no" is not working, that is the time to escalate to your manager.

Melissa once worked with a stakeholder who was a project manager for the operations team. The stakeholder was not happy with the way one of Melissa's PMs presented the roadmap, which was in roughly a "Now, Next, Later" format. This stakeholder asked the PM to add dates, and they said no, so the stakeholder created her own version of the roadmap, now with dates.

The problem was that when the stakeholder created her own version, she not only put unrealistic dates (such as condensing a quarter's worth of work into a month), but she also added her own new ideas into the roadmap, instead of sticking with what she originally agreed to. This "alternate-reality roadmap" was then shown to other stakeholders, and worse, was presented as the PM's work.

Melissa's PM alerted her to the situation and the two of them met with the stakeholder. They approached her with curiosity and asked a number of questions, such as "What were you asked to present?" and "What caused you to put together your own version of the roadmap?" The stakeholder claimed that *her* stakeholders (leaders of various operational functions), were asking *her* for dates. Melissa spoke directly with the other operational leaders and discovered that they were asking for dates but were also happy with the "Now, Next, Later" format.

Melissa and the PM went back to the stakeholder and suggested that the PM present the roadmap directly to the operational leaders, with the stakeholder there. That way they wouldn't have to change their process, and nothing would be lost in translation.

Disagree and Commit

Sometimes agreement among all parties is impossible. In these situations, you must decide whether to keep discussing, or to make a decision and move forward. The difference between agreement and alignment is that you can be aligned on a path forward without necessarily agreeing. Amazon calls this "disagree and commit," which is when someone disagrees, expresses their objection, but then commits to the decision anyway. Someone might "pick their battles" by committing to a decision to remove a roadblock.

How you think about "disagree and commit" depends on your role in the disagreement. One situation is where your stakeholders don't all agree with each other, or with the decision you've made. In this case, you should work one-on-one with the dissenting stakeholders to identify their concerns and align on the decision, even though it may not have been their first choice. For example, Melissa once worked with an engineering lead who was not on board with a decision to prioritize a customer request. Melissa volunteered to take all the blame if it didn't work out well, and told the engineer he could send any stakeholder questions to her. The engineering lead decided this was enough for him to commit to the decision, even though he disagreed.

Another "disagree and commit" situation is when you, as the product manager, commit to a decision your stakeholders are aligned on, even if you don't agree with it. For example, we know a director of product who was given a mandate by his COO to shift his team's focus to a new initiative, but he didn't think it was a good idea. All of his other stakeholders were also on board with this plan. The product director talked through his concerns with his stakeholders and the COO, but they would not budge. He went back to his team and they found a way to use half the team's resources for the new initiative and retain the other half for pieces of the existing roadmap. Even though he disagreed with the COO, the product director committed to the new initiative anyway, and it turned out to drive more value than he initially thought it would.

Irie Says No

Irie decides to propose a few of Sparks's requests to do now, and to attempt to say "no" to the rest by "registering them for later." She goes through his list and identifies three requests that meet the product objectives and research findings."

Christina says that two of the three look like they are relatively small, and closely aligned with work already in progress. Engineering says they can accommodate these two items. They might be able to take on the third item near the end of the quarter. Irie takes the news back to Sparks.

"Only two?" asks Sparks. "Not good enough."

"I can't make resources appear out of thin air. Realistically," Irie adds, "to do more than this we would have to postpone most of the rest of the roadmap until next year, or even the year after. I don't think that would be the right thing to do for our objectives."

"That's where the partnerships can help," says Sparks. "Just trust me. Put together a roadmap with all these features this year, and maybe our resource problems will go away."

"Well," says Irie. "It only makes sense to put on the roadmap what we think we can actually do with the resources we have." Irie pauses and then proposes an alternative idea. "But what I can do for you is to create different versions of the timeline for different resourcing levels."

Sparks pauses to consider this. "Yeah, let's do that, that would be helpful...Actually, that might be all I needed in the first place. Can you put together a roadmap with all these requests and tell me how many resources you'd need to get it all done in, say, a year?"

"I can try to do that," says Irie, "but it's not going to just list out all your requests. I'm going to group them into themes and put those on the roadmap. I'm not 100% convinced that your specific list is the right way to solve all these problems, but as I was looking over the list, there were obvious patterns of problems they're intended to solve."

"But they've asked for those specific features," says Sparks. "VigGuard and Alvex want to know how long it will take for us to build them all."

"I know that's what they said they want," says Irie. "But maybe we can talk with them and see what's behind it. If we go into detailed estimation for all these things they asked for—estimates that do require engineering resources to make—we will distract the team from delivering what we've already committed to. That would make us look like an unreliable partner."

"True," says Sparks. "But I can't bring you in on those conversations with them just yet."

"I understand," says Irie. "If we can show them we can deliver on our commitments, perhaps they will trust that we can do the other features later, or even sooner if we add the right resources."

"Your roadmap went over pretty well with Liz and the other execs," admits Sparks. "I guess I'll have to rely on you here."

"Thank you," says Irie, pleased to finally have a small win with Sparks. ■

No.

Takeaways

There will always be changes you'll need to account for in your product. Rather than resisting change, embrace it and work it into your process. Here's how to manage roadmap updates:

- A clear intake process will help you manage stakeholder requests by consolidating them to remove distractions. An intake process should include four steps: 1) submit requests, 2) triage requests, 3) follow up, and 4) decide on a plan.
- Roadmaps must change as the environment changes. Use a roadmap planning cycle to assess learnings and reevaluate your plans. Ensure you start early enough to draft the new roadmap and iterate with stakeholders.
- Avoid Alignment Decay by reviewing your roadmap periodically with stakeholders to ensure continued alignment.
- Utilize shuttle diplomacy by hold pre-alignment meetings to address stakeholder concerns individually.
- Use decision logs to document your decisions for later reference.
- Ideally you and your stakeholder will arrive at an answer of "no" together. If that's not possible, use what you know about your stakeholder to craft a narrative that will help them understand why "no" is the right answer.
- Escalate when needed to ensure your stakeholder hears a strong "no" now, rather than a string of "maybes," disappointing them in the end.
- Sometimes you or your stakeholders need to "disagree and commit" to remove a roadblock that impedes progress.

There is almost always something you don't know

Chapter 7

Challenges

Despite your best efforts, sometimes you have to work with a stakeholder who defies logic. When you are met with the biggest stakeholder challenges, you have to ask some serious questions and make some serious decisions.

In this chapter you will learn how to do the following:

- Check yourself and the context to verify that your stakeholder is truly difficult.
- Set relationship goals to deal with difficult stakeholders.
- Evaluate your future options.
- Set quitting criteria to decide how to move forward.

Irie has been having a rough time trying to manage Sparks. We'll join Irie as she's once again tricked into thinking she's making progress.

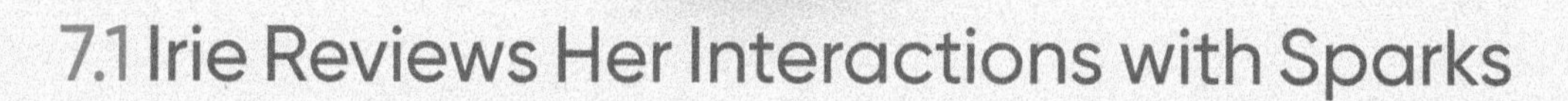

7.1 Irie Reviews Her Interactions with Sparks

"I like this version of the roadmap," says Sparks, looking at one of the slides Irie is sharing. She has prepared, as agreed, a few different versions of the high-level timeline for different resourcing levels. They usually meet in the office, but today Sparks is traveling, so they are meeting via video. Sparks looks like he's sitting at a desk in a hotel room. "It gets us everything they asked for. Great job."

"I like it, too," replies Irie, hesitant. "But it'll require at least two more teams working in parallel."

Sparks ignores this and asks if Irie will present this version of the roadmap to their potential partners. Irie protests that this was just for them to project the needed resources. "It is strictly internal. I can't commit to any of this publicly, and I'm sure Sri would agree," she says.

"If you had the resources, you could commit, right?" he asks.

Irie hesitates, then nods. Sparks continues, saying, "We'll be talking to people who can help us with that. We'll tell them the roadmap is contingent on their help."

That sounds reasonable to Irie, but she finds she is still uncomfortable. "I'd feel better about checking this direction with Liz," she says. "It feels off the main path she laid out toward our vision and I'd feel better if I got her okay."

"We really need to give Liz the space to get better," he says. "I don't want you bothering her with any of this."

"But this new theme on lowering medical costs isn't supported by our user research," Irie points out. "Most of our customers are covered by insurance, so they care more about rising premiums than about out-of-pocket costs. And a lot of these other themes aren't yet validated. It feels like we're trying to inject AI everywhere we can. It's not customer-driven. I really think Liz—"

"Send me the slides so I can add a few other specifics to the deliverables," Sparks says, interrupting. "I've got to jump on another call." And with that, he disconnects.

Irie finds herself alone at her desk, frustrated. Over and over, Sparks seems to see reason and then revert to old habits without warning.

Irie runs through all of her interactions with him since their first meeting, where he tried to insert the AI assistant into the product priorities. She gave him respect because he is a Power Player despite a lack of clarity around his authority. She tracked him down and established some level of rapport over coffee. She approached him using his Dominant thinking style instead of overwhelming him with data. She thought she had established some amount of trust with him, but even after she got him to agree to product objectives, he still tried to add unrelated things.

It might just have been his obsession with AI, she thinks. Except now he's adding other things to the roadmap as well, like the cost reduction theme. As soon as they have a roadmap, he wants to change it. And when she introduces a process for changes, he demands she go around it with this secret alternate version.

Worst of all, every time she points out inconsistencies, asks for reasons why, or wants to check with Liz, he puts her off. *I can't work like this,* she thinks. She feels overwhelmed and out of control in ways she never felt when she worked for Darius.

She decides to see if Darius has any advice, and they agree to meet later at his favorite spot for coffee. It's a small independent café a short distance from the Helthex office. Walking to the café, she realizes how stiff her legs and shoulders are, probably due to the stress of dealing with Sparks. The walk helps her work out the kinks but it also makes her realize how tense she has become.

When she arrives at the café, Darius is listening to the woman behind the counter describe her challenges sourcing quality coffee. Darius is listening sympathetically while the woman makes his drink. Irie greets Darius, and he asks, "Doesn't your cousin grow coffee?"

Irie gets the shop owner's card and promises to connect her with her cousin on the Island. "He grows the most flavorful beans you've ever tasted," she promises. "I'll bring you the bag he sent me last week."

She and Darius sit with their coffee and she recounts recent events, asking if he has any advice. Summarizing what he's heard from Irie, Darius says, "The company is short on resources, VC funding is not yet in the works, your CEO is out of commission, and the other founder is pushing some secret hurry-up integrations that he sees as a fix to the resource problem. Sounds like acquisition talks to me."

"Really?" says Irie.

"If these were just ordinary integrations, they wouldn't be secret," Darius says. "And if these partners were just offering small investments or lending you some of their people, he probably would have told you. Actually, it's a good sign that he is simply withholding information. I've known execs who lied outright to their people about things like this to protect their own interests."

Irie is skeptical about Sparks's motives. "He opened up to me about his father's financial challenges. He even told me how he hated his real name and why he goes by that nickname. But now I think it was just a bunch of BS to make me feel like I was getting the real story. He didn't tell me he was planning to sell the company. And I think he's hiding it from Liz, too. She seems happy to talk to me on her own," she adds, "but he tells me not to share any of this with her."

Darius reminds her that they are speculating about these things, so she should not assume anything until she has more information. "We don't know the whole story," he says. "It's best to assume positive intent until proven otherwise."

"This is so frustrating," Irie confides. "I feel like I'm operating in the dark. I can't make good product decisions if I don't have the whole picture." She pauses for a moment to gather her thoughts, then adds, "It's worse even than that, though. How can I continue working with Sparks if I can't trust him?" she asks.

Darius says he's developed a sort of rubric for dealing with difficult stakeholders. "As an advisor, I run into all sorts of people and situations," he says. "Most people don't mean to be difficult. Either they are so preoccupied with their own issues they don't realize how they come off, or there are things they really can't share and they try to make it sound as good as they can."

"In fact," he adds, "I find that when someone's behavior doesn't make sense, there is almost always something you don't know. Maybe it's about them; maybe it's about the situation. Sometimes it's even about yourself and how you are reacting in the moment. Let's talk about how to assess the situation and some techniques I've found helpful for dealing with difficult people. Maybe this is something to add to your Product Playbook." ■

When someone's behavior doesn't make sense, there is almost always something you don't know.

Dealing with Difficult People

Clearly, Sparks is a difficult person for Irie to deal with. We all have a mix of easy-breezy stakeholders and difficult ones. (Some of us even have our own Sparks!) When you're evaluating the best ways to improve your relationship with a difficult stakeholder, consider that sometimes it's not really about them at all. To decide whether your stakeholder is simply a difficult person or whether there's something else going on, you have to first ask yourself two key questions: "Is it me?" and "Is it the situation?"

Check yourself first

Some people's first instinct is to blame relationship problems on the other person. Other people's first instinct is to blame a bad relationship on themselves. The best thing to do is to consider both possibilities, because it's more than likely a combination of the two.

To assess if *you're* the problem in the relationship, you need to check your biases. There is a concept in psychology called "naïve realism," which is the human tendency to believe that one's own view of the world is based on fact, and is free of bias. This tendency drives us to believe that if other people don't agree with us, they must be uninformed or irrational. We must be right and they must be wrong. Taking a more empathetic approach, we might realize that there is in fact no "wrong" at all, just different perspectives. To combat naive realism, start with the assumption that you're wrong, and brainstorm alternative realities.

Another way to check yourself is to get a sense for whether other people at your company also have a challenging relationship with this stakeholder. Is it just you, or does everyone have a problem with them? If everyone else has good relationships with the stakeholder, try to understand what they're doing differently. Maybe there are some tips you can learn for improving the relationship.

Sometimes you may actually find that you're emulating the bad behavior you dislike in your stakeholder. This can happen easily in a culture that is particularly stressful, not psychologically safe, or rewards bad behavior. Take a look at your own behavior and decide if you're possibly part of the problem. If you're acting in a way you wouldn't normally act, you could be making the situation worse.

And finally, check in on your own physical and mental state. If you've had a counterproductive conversation with a key stakeholder, could it be that you were "off" that day? Is there something causing you stress? Did you forget to eat lunch? Are you tired or under the weather? If you find that your own behavior has contributed to a bad interaction, apologizing shows humility and can make you more relatable. Taking responsibility may also lead your stakeholder to make an admission of their own. And maybe the two of you can "start over."

Check the context

If you find that *you* are not the cause of someone's bad behavior, another environmental factor could be at fault. Is there anything happening right now in your organization that might make your stakeholder stressed? Have their incentives changed? Is their manager making their life difficult? Are they dealing with a stressful situation at home? Finding out more about the context can help explain the situation, and showing empathy about it can help improve your relationship.

Following from naive realism, there is another psychological phenomenon called the "fundamental attribution error." When thinking about someone's behavior, this bias makes us overemphasize personality traits and under-emphasize situational or environmental factors. In other words, if someone is late, we tend to assume it's because they're lazy, not because they got stuck in traffic. Ask yourself if there are any environmental factors that could explain problems in your relationship with your stakeholder.

Melissa knows a product manager who was working on an important product launch. His marketing stakeholder, who was supposed to be his partner in the launch, was frequently late to meetings or missed them altogether. When he did show up, he was unprepared, distracted, and even rude sometimes. The product manager took the time to ask the marketing stakeholder if everything was okay at home. The marketing stakeholder admitted that his wife had been in and out of the hospital with an illness, which is why he had to miss meetings sometimes. The product manager was empathetic with the situation and said he could be very flexible, asking only that the stakeholder inform him when he needed to reschedule a meeting. This helped them create a much better working relationship.

Truly difficult people

When it's not you, and it's not the environment, your stakeholder might just be a difficult person. In this case, you can try to discover their intrinsic motivations to see if you can turn a negative relationship into a positive one. According to Tony Fadell, in his book *Build* (Harper Business, 2022), there are "several different kinds of assholes." Some assholes are just mean, but some are well meaning and don't realize how their behavior affects people. These people he calls "mission-driven assholes."

"Mission-driven assholes" may simply be very passionate about a particular issue, and may not see their behavior as "difficult," rather as simply the best way to get results. Fadell explains: "Unlike true assholes, they care. They give a damn. They listen. They work incredibly hard and push their team to be better—often against their will. They are unrelenting when they know they're right, but are open to changing their minds and will praise other people's efforts if they're genuinely great."

With truly difficult people like these, you often need to invest extra effort to build good relationships. Give them additional attention, understand what drives them, and acknowledge their point of view—even publicly, and even when you disagree. Mission-driven people want to know that you "get it." Once they are comfortable that you do, you can have a much more reasonable conversation.

Finding the time and energy for this can be draining, especially since it means spending extra time with someone you may not get along with. When these people are critical to your plans, though, the investment of time in making them from a Challenger to an Ally* is usually worth it.

* See "Moving Power Players Toward Alignment" in Chapter 1.

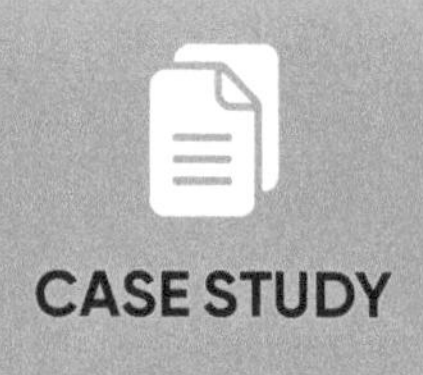

CASE STUDY

Hearing Bruce's Architect

Bruce once worked with a brilliant software architect. She understood the technical underpinnings of the product better than anyone else, but she was frequently difficult in team meetings. New ideas set her off every time. Anything that didn't fit within her existing schema was dangerous.

She disrupted several planning sessions in a row before Bruce decided to invite her to talk over lunch. There he explained that he was concerned about confusing the team with open debate between the two of them.

She was grateful for the opportunity to talk, sharing her own frustration with expressing herself. Vigorous debate was part of her upbringing and she complained that most people seemed unwilling to engage with her the way she was used to.

Bruce valued her input and so he proposed they have these debates one-on-one, rather than with the whole team. Going forward, he would come to her first with new ideas, hash out the issues, and then bring an aligned approach to the rest of the team.

This bargain proved its usefulness in the very next meeting. Bruce summarized the proposed direction and then acknowledged the risks the architect had raised with him in their one-on-one the day before. He framed her position as "concerns" to be addressed rather than "objections," wording he had previewed with her.

She supported the general direction, adding a few specific issues to be worked through with the team as a whole. She'd been heard and her concerns had been translated* into a form the team could handle. It was extra work, meeting with her separately, but well worth it to get the best out of the whole team, including the team's most brilliant member.

* As product managers, we often find that our jobs include "translating" between different functions, e.g., communicating technical problems to marketing leaders. But it's also helpful to be able to "translate" among different types of communicators, even within the same function, as Bruce did in this example.

Set Relationship Goals

In her book *Getting Along* (Harvard Business Review Press, 2022), Amy Gallo shares a useful technique for dealing with especially difficult people: set relationship goals. Gallo says that "identifying your goal will help you avoid getting pulled into any drama and stay focused on constructive tactics."

Write down your relationship goals and pick the top two or three to focus on. Here are some ideas for goals you might want to set for dealing with your difficult stakeholder:

- Align on a decision to move your project forward.
- Create a healthier working relationship that you can live with for the foreseeable future.
- Simply tolerate each other without getting into arguments.
- Feel less angry or frustrated in meetings with them.
- Reduce the total amount of time you have to spend with the stakeholder, while still getting your work done.
- Get yourself moved onto a different product so you don't have to work with your stakeholder anymore.
- Stop focusing on them outside of work.

When setting relationship goals, beware of subconscious goals that will distract you from improving the relationship. These may make you feel better for a while, but ultimately trying to get revenge isn't a helpful goal. Here are some common subconscious goals to avoid:

- Get them fired or make them want to quit.
- Make them feel miserable, embarrassed, frustrated, etc.
- Expose their bad behaviors to the rest of the company.

Focusing only on what you can control, and letting go of what you can't, may help relieve some anxiety about the situation. As Adam Grant says in his book *Think Again* (Viking, 2021), "I no longer believe it's my place to change anyone's mind. All I can do is try to understand their thinking and ask if they're open to some rethinking. The rest is up to them."

Take Care of Yourself

If you've done everything you can but working with this stakeholder is still a nightmare, you may need to step back and focus on yourself. It's not worth risking your physical or mental health for this person. Take a deep breath and remember that it's only a job, it's not your whole life.

It may be worth enlisting your manager to help. Assuming they're aware of the situation, and you've been sharing updates with them, they will hopefully be empathetic. (If they are not, that's a whole other story.) It may be worth taking some time off to de-stress, if you can. It may also be valuable to seek help from a mental health specialist. If the situation becomes untenable, you may want to consider other options like starting to look for a new job.

Escalate

Before you think about quitting, however, you should consider escalating the problem to someone higher up who may be able to help. Many people think that escalating makes them look weak. But in reality, people who escalate problems instead of letting them fester save a lot of time and headache, often avoiding unnecessary delay, even attrition. And you may discover you are not the first person to bring forward a problem with this difficult stakeholder. Keep in mind that the best kinds of escalations come not only with a problem statement but also with one or more proposed solutions. Having a solution in mind makes you look like a creative thinker rather than just a complainer.

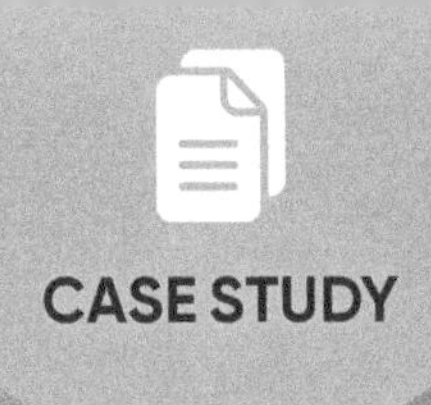

What Gabriel Didn't Know

When Gabriel started at his company as head of engineering, he had a great relationship with his boss, the company's CTO. After a few months, though, the CTO's behavior began to change. He accused Gabriel openly of disloyalty and deviousness. This was sometimes triggered when they disagreed, but sometimes it came out of the blue. The CTO began having unprofessional outbursts of swearing and name calling regularly.

At first, Gabriel thought this was about him in some way, but feedback came from others around the company that his boss's disruptive behavior was indiscriminate. People from every department began avoiding the CTO, fearing his ire.

Gabriel approached his boss empathetically, asking if he was okay. The CTO said he knew he had a hot temper, apologized, and promised it wouldn't happen again. But it did.

Gabriel took him out for coffee, just to get beyond the work bubble. Same result. He tried everything. Eventually, he went to the CEO to suggest his boss needed help.

The CEO was aware of the problem. The CTO was brilliant, he said: the primary innovating force in the company. He couldn't afford to lose him, but he also couldn't afford to allow his behavior to alienate everyone else. Professional help seemed worth trying, so the CEO hired him an executive coach.

He also asked the CTO—a great technologist—to leave management tasks to Gabriel. This was fortuitous as the coach soon helped the CTO see that he was acting out of frustration with the growing demands of management. He longed to return to the technical work he loved.

You may feel that most people are unable or unwilling to change, but in this case, adjusting the job to fit the person worked well. Relieved of management headaches, he returned to being a respected member of the team.

Gabriel was patient because the CEO had taken action. He set a deadline, agreeing with himself that if things didn't improve in six months, he would start looking. It took less time than that to see changes, though, and Gabriel soon enjoyed working with his former mentor again.

Irie Sets Relationship Goals for Sparks

After leaving the coffee shop, Irie thinks about Darius's advice on her walk back to the office. She agrees she needs to widen her focus from how difficult Sparks can be to look at herself and the larger context. She decides to start by speaking with Sri, as she has come to trust him implicitly.

She finds Sri in his office with Yacob, and explains the situation. "Well, it's not you," Sri says without hesitation. "Sparks has always been hard to deal with, even back when he was CTO."

"Sparks was CTO?" Irie asks, incredulous.

"He and Liz started the company," Sri says.

"He wrote the first version of the app," adds Yacob. "It was one of the first to grab health data from multiple different sources and present it all together."

"When did they hire you?" Irie asks Sri.

"A couple of years later, after they'd built out the engineering team to twenty or so," Sri explains. "Yacob was here before me as a lead engineer."

"Reporting to Sparks was…not working for me," Yacob says. "Priorities changed every week. He prototyped things over the weekend and then threw it to us to make sense of it. People were stressed out and ready to leave."

"Sounds familiar," says Irie.

Yacob continues. "Liz finally decided to hire Sri to run engineering when it became obvious Sparks wasn't the manager type. Sri's done a great job, but it feels like we're back there now with Sparks jamming stuff in through product," he adds, reminding her of the product update meeting on her first day. "I was frankly hoping you could help shield us from him, but I'm not sure anyone could."

"If it's not me," Irie says, "then maybe it's about these partnerships. Darius thinks there is more going on here than partnership, like maybe these companies want to acquire us."

Sri and Yacob exchange glances. Clearly, this has occurred to them, too.

Sri closes the door and sits back down before continuing, "Philippe asked me confidentially for a list of possible staff cuts. We're short on cash and, with her illness, Liz hasn't been able to fundraise."

"Oh, that's awful," Irie says. Simultaneously, she is thinking this confirms what Darius suggested about VC funding.

"Both of you are safe," Sri adds quickly. "I can't see us managing without you, but we'll have to make some pretty deep cuts to both of your teams." Sri pauses to see how Irie and Yacob react. They are still processing as he continues. "I don't have any more facts than that," he adds, "but a buyout could help keep us afloat."

"Do you think Liz knows?" Irie asks.

Sri thinks for a moment. "I would normally assume yes," he says. "She's the CEO and she and Sparks have alway been very close."

"As close as Sparks can be to anybody," Yacob says.

Sri ignores Yacob's sarcasm and continues. "But Liz is going through a lot. It's possible he is keeping this from her to protect her, because he thinks he knows best."

Later that day, Irie gets similar feedback from Ella and Liandri. They reassure her she is doing all she can and encourage her to take a day off to get some distance from Sparks. "When I get fed up with customers complaining, I go for a long hike with my dogs," says Liandri. "It clears out all the gunk."

Neither of them knows any more than Sri about these partnerships, but Ella volunteers to do some digging about the companies. "When you've been in sales as long as I have, you develop a pretty extensive network," she says.

That evening, Irie decides to curl up with reruns of *Star Trek*. Something about the naive optimism of Kirk and Spock flying around the galaxy fixing broken civilizations has always comforted her.

After half an hour, though, she finds she is still thinking about Sparks. She gives up, turns off the TV, and begins writing relationship goals for them.

She writes:

Get Sparks to go along with the roadmapping process

Then she almost crosses it off. *What's the point of trying*, she asks herself. *He's impossible.* Then she writes:

~~Get Sparks fired~~

She immediately crosses this one off. *Tempting but not constructive*, she thinks. She writes instead:

Avoid dwelling on Sparks outside of work

Thoughtfully, she then adds:

Help Sparks become aware of how hiding his plans is bad for the company

She reviews her goals with satisfaction. *The last one seems the hardest*, she thinks, so she resolves to sleep on it after she works on goal number two with some more *Star Trek*. ■

7.2 Irie Considers Quitting

Sparks is still traveling the following week. She receives two meeting invitations from him for the week after, each titled "Roadmap Presentation," but he doesn't respond to her messages asking for details.

She shares her frustration at a meeting with Sri, Ella, Liandri, Sergey, Yacob, Divya, and José. Ella then reveals what she has learned about their potential partners. Alvex has spent a lot of money developing their chatbot technology and is investing in social media advertising and airport signage. "But it's all showy tech without practical uses for it that make money. They look at us as a concrete application of AI chat," Ella summarizes.

"This is interesting," Irie says, "but we already knew that they wanted to partner with us on their chatbot, from what Sparks said."

"Right," Ella says, "but Darius was spot on thinking they wanted to acquire us. I talked to a friend of a friend on their board and they need more than a channel to embed their tech into. They actually want to embed *our* stuff into *their* app."

Ella responds to puzzled looks around the room by explaining that they envision an "everything app" powered by an AI that can advise on anything, including health.

Sri says that doesn't sound bad, but then Ella reveals that they are only interested in the integrations Helthex has built. "They plan to scrap everything else and just let the AI figure out what to recommend from public sources and user data."

"That is a horrible idea!" says Sergey. "The internet is full of bad advice."

"And we know it won't work," says Divya. "We tested with their chatbot. It will require months of extra work building on our stuff and theirs. Throwing that away would be going backward."

Ella adds that they plan to keep the engineering team but that they will eliminate customer support altogether. "They think the AI can handle it," she explains.

Fuming, Liandri says, "I wish them luck with that."

While they are still absorbing this news, Ella relates what she's learned about VigGuard, a large US health insurance carrier. They see the opportunity to use fine-grained user data to pinpoint people with unhealthy habits for specific programs.

"Like advice on how to lose weight or to get better sleep," says Sergey. "We do that already."

"I know someone on their sales team, though, who had a little more information," Ella explains. Apparently, recent changes in US law permit them to raise rates on people they deem higher risk and even drop them from their plans if they are unprofitable. "And once one insurance company drops you, none of the others will touch you."

Looking at the shocked faces around the table, Ella says, "I've worked for companies like these before. I don't think anyone there is actively evil. I just think the incentives inevitably push them toward what makes money."

"I can't believe this is what Liz would want," Irie says. "Either option would mean abandoning her vision."

"This is horrible," says Sergey. "I will not work for such a company."

"This is not why I joined Helthex," says Liandri, disgusted, "but at least with VigGuard my team can keep their jobs."

Sri is not pleased either. "Maybe we can be a positive influence on whoever acquires us," he offers.

"It may not matter what they want to do with the data, actually," José volunteers. "I segmented the usage data and I found out that our corporate users are hardly using the app. It's the same with our VigGuard pilot group: one-tenth the usage of our B2C customers. The good news," he adds, "is that the averages have been fooling us. People who sign up on their own are more engaged than we thought. They really value the app and corporate deals are just bringing down the numbers."

Discouraged, the group breaks up without any decisions or action items. Irie and Sri stay behind. "I worked at a company that was acquired before I left," she explains. "They said they wanted us to re-energize innovation in their company. We were all excited. But it eventually became clear that nothing was going to change. A lot of my friends left disappointed. Some stayed because they needed the job, of course, but at this point everything we built is pretty much gone. I came here because I believe in Liz's vision, but now..." She finds herself staring out the conference room window.

"I get it," Sri says empathetically. "Staying or going is a personal decision. But the information we have is all second- or third-hand," he adds. "And we need you leading product here. I hope you'll consider staying." Irie is silent. Sri adds, "If it helps, I watched this great video on when to quit your job. I'll send it to you."*

This offer shakes her out of her reverie. She's not ready to quit, she realizes. She'd rather develop a plan to make things work than give up now assuming they won't. Sri texts her the link to the recording and the words "when to walk away" catch her eye. ■

* *https://youtu.be/50wcc7L232U*

When to Quit

If you're thinking about quitting, you should think about the decision as two sides of a coin. Just like saying "no" can be reframed as saying "Do you want this or that?" then the decision to quit is actually "Do you want to quit or to stay?" Since you're already on the "staying" path, you have to decide whether to continue on your current path, or change paths. And delaying the decision to quit is simply deciding to stay for a bit longer.

You should decide to quit when staying where you are stops being the best choice for you. As Annie Duke says in her phenomenal book, *Quit* (Portfolio, 2022), "If you quit something that's no longer worth pursuing, that's not a failure. That's a success."

Why it's hard to quit

There are many reasons why our irrational brain has trouble giving up on our current path (Figure 7-1), especially "in the moment." By understanding how our brains work, we can start to recognize when we're making decisions that are perhaps irrational.

Figure 7-1. Psychological principles that make it hard to quit

Principle	Description	Why this is important
Sunk cost fallacy[*]	We tend to consider money/time already invested when making a decision of whether to continue spending more.	Sunk cost is not related to future expected value. Considering sunk costs usually causes people to "throw good money after bad" and stick longer than they should.
Loss aversion[†]	Losing feels about twice as bad to us as winning feels good.	People tend to quit when they're ahead and stick when they're behind. We should decide whether to quit or stick based on expected future value, not whether we're currently winning or losing.
Data inequity[‡]	Staying feels more comfortable than quitting because you can measure your current situation, but you can only predict your future situation.	We tend to stick with what we know rather than trying something new. "The enemy you know is better than the enemy you don't."
Endowment effect[§]	We value an item we own more than an identical item we don't own.	We may think our current job has a higher value than another, roughly equivalent job (even though the new job would not have our troublesome stakeholder in it).
Omission-commission bias	We are more concerned with outcomes of things we actively do (commissions) than we are with outcomes of things we let happen by not acting (omissions).	We tend to stick with the status quo because we worry that a mistake "we made" is worse than a mistake "that happened to us."
Cognitive dissonance	When new information contradicts what you believe, you tend to discount the new information rather than change what you believe.	If our situation at work changes, we might not accept it for a while and ignore signs that we should quit, thus staying too long.
Escalation of commitment[¶]	As we near a goal or milestone, we tend to increase our desire to continue toward the goal, even if the goal becomes unattainable.	If we are working toward a milestone, like launching a product, we may stay longer than is good for us because it feels like a waste to quit right before the end.

* Richard Thaler, "Toward a Positive Theory of Consumer Choice," *Journal of Economic Behavior & Organization*, Vol. 1, 1980.

† Daniel Kahneman and Amos Tversky, "Prospect Theory: An Analysis of Decision Under Risk," *Econometrica*, Vol. 47, No. 2, March 1979,

‡ Duke, *Quit* (Portfolio, 2022),

§ Thaler, "Toward a Positive Theory."

¶ Duke, *Quit*.

How to Decide Whether to Quit or Stay

With all of these normal human decision-making foibles working against us, we need a framework for making more rational decisions. We tend to make even worse decisions "in the moment," so setting up the criteria for our decision making in advance is the best way to combat our irrational brains.

Think in expected value

One thing you can do is think about your decision in terms of "expected value." In plain terms, this means that you should think about what you expect to happen in the future, and compare your two choices. For example, if you're miserable now, and you can't change things, then the expected value of being miserable in your current job a year from now is 100%. Despite the uncertainty of switching jobs, the expected value of being miserable a year from now if you quit is something less than 100%. In that case, the logical thing to do is to switch jobs.

Set quitting criteria

To prevent our irrational brains from taking over during a decision, it's helpful to decide in advance what would make us quit or stay, and write that down to refer to if the situation arises. In *Quit*, Duke calls this setting "kill criteria." She says that kill criteria are critical to allow you to cut your losses. "When you set out clear kill criteria in advance and make a precommitment to walk away when you see those signals, you are just more likely to follow through, even when you are losing."

For example, you may decide that your quitting criteria is that if you can't get your stakeholder to include you in critical meetings that affect your success by March, then you'll start to look for a new job. Note that the example included a criterion, a date by which it needs to be met, and what happens if it is not met.

Deciding to quit doesn't mean you have to turn in your notice at that very moment. It could mean that you start actively looking for a new role so that you can change jobs when you get an offer.

Here's how to set quitting criteria:

1. Think about your goals, including your stakeholder relationship goals. Decide what success looks like and turn that into measurable objectives. Write it down.
2. Find a buddy who can review your quitting criteria and help keep you to your word. This should be someone you trust to be honest with you, like a spouse, a good friend, or a mentor.
3. Decide when you want to make the decision to quit or stay: set that as a date to evaluate your quitting criteria.
4. Make an appointment with your quitting buddy. You can also meet with them periodically to see how things are going along the way.
5. It can be helpful to imagine in advance how that quitting-decision conversation is going to go, to get used to the idea of making a big and potentially scary decision.
6. On your set decision date, look objectively at your quitting criteria: were they met or not? If they have not been met, you should quit (or stay and start actively looking for a new job).

Irie Sets Quitting Criteria

Reviewing her notes from the video, Irie decides she still isn't ready to quit, but the idea of making her criteria for staying or not staying crisp and clear appeals to the analytical side of her.

She writes out the key reasons why she took this job, the "expected value" she saw at the time.

Reasons for joining Helthex

- Opportunity to contribute to Liz's vision
- Opportunity to lead product management at a growth-stage company
- Opportunity to work with and learn from great people

She reflects on the usual "value" aspects of a job such as the pay and benefits and decides not to add them. She actually took a pay cut to join Helthex because of the reasons she listed.

Deriving quitting criteria from these expected values seems straightforward to her.

Quitting criteria

- If the company abandons Liz's vision, she should quit
- If she is no longer leading product management, she should quit
- If she no longer works with great people, she should quit

Looking at these criteria, she feels they are all at risk. If Ella's intel on their potential acquirers is right, their vision is very likely dead. And she would probably end up reporting to someone in product management at the parent company rather than Sri. And if everyone she's come to enjoy working with at Helthex leaves like they did after her prior company was acquired, the last of her criteria would be checked off as well.

On a roll now, she feels that Sparks has already undermined her "great people" value. And since he seems to be dictating the roadmap, she questions whether she can accurately claim to be leading product management.

She decides she needs to add one of her Sparks relationship goals to her quitting criteria. If she can't convince him to be open with her, Liz, and the rest of the leadership team about what's really going on, she will quit.

Per the video, though, she decides she needs to give things time to play out. She feels she will have solid evidence on her first three criteria within the next six months. Getting Sparks to come clean with his plans has to happen more quickly, though, she decides. She resolves to ask for help from the rest of her product council.

Sparks is still absent at the product council's next meeting, but Philippe joins the call remotely and addresses the possible acquisition right away. He confirms with each member of the group when they will present and what they will cover at the upcoming meetings with Alvex and VigGuard. Everyone agrees, but Irie sees clearly that no one is happy, not even Philippe.

“Philippe,” Irie asks, “why are we doing this?”

Philippe apologizes for not letting everyone in on the details earlier, but explains that either of the deals will help them avoid deep cuts that would put the future of the company at risk. “I can’t be responsible for letting the company die, not while there is a chance of keeping Liz’s vision alive.”

Ella recounts to Philippe what she’s learned from her contacts about these companies’ true intentions, but Philippe just raises his hands into the camera’s view in a helpless gesture. “The numbers are the numbers,” he says.

“Well, I don’t want to work for a company that hides behind the numbers,” Ella says.

“And I think I’m likely to lose both Sergey and Liandri,” she adds.

Sri says, “I’m with them, Philippe. When we first talked about this, I was thinking about saving my people’s jobs, but word is getting out and I’m seeing a lot of updates to online profiles. We’re going to lose a lot of great people.”

Listening to her teammates, Irie realizes they all have the same quitting criteria she does. “There must be another way,” she says. “We all have too much at stake.”

Philippe shares his projections and the numbers are indeed bleak. Without cuts they will be unable to make payroll within four months. “If we cut headcount by 30% across the board, we can make it to the end of the year, but unless we have a revenue miracle, we’re still out of cash.”

“I thought Liz was going to raise money before the end of the year,” says Sri.

“That was the plan, yes,” says Philippe, “but she has been out so much that I don’t see that happening.”

“Maybe if we talk to Sparks and Liz together,” Irie proposes, “we can figure something out. Sparks may not care about all of this, and I don’t know how much Liz knows about it, but we have to try. This is our company, too.” ■

7.3 Irie Puts Helthex to the Test

Irie agrees to reach out to Liz and Sparks to set up a meeting with them and her product council. When she gets back to her desk, though, she finds she already has an invitation to meet with the founders in her inbox. The time is set for later that day, and she is the only invitee. She is apprehensive about the nature of the meeting but decides to seize the opportunity to speak with them both and perhaps bring the rest of the team together later.

She is determined to be transparent with Sparks and Liz about what she and the team have learned or deduced, as well as how strongly they all feel that selling to these companies would be a mistake. What happens after that, she is not sure.

Irie arrives in Sparks's office to discover he is already speaking to Liz on his monitor. The image, clearly from a phone camera held in an unsteady hand, is of a woman in a hospital bed, looking thin and drawn, but smiling through fatigue. Tubes and wires disappear into her hospital gown and a thin knit cap covers her hairless head.

"Irie, I'm so glad you're here," Liz says, her voice still strong. "Sparks has told me about the great work you've done these last few weeks. I'm so happy he has your help."

Clearly, Liz's health problems have worsened. Irie isn't sure how to react or what she is allowed to ask, so she simply thanks Liz.

"Irie," Liz begins again, "as you can see, my situation has accelerated. I'm optimistic about the long term, but the doctors tell me I am going to be unable to work for a few months while I fight this thing."

Listening, Irie finds she is holding her breath. She makes a conscious effort to breathe out slowly, calming her body.

"I was hoping this wouldn't be necessary, but I'm taking a leave of absence and I'm making Sparks interim CEO for the duration," Liz explains.

Numbly, Irie visualizes the note she wrote about quitting criteria. It never occurred to her that Sparks might become CEO or that she might have to report to him, even indirectly through Sri. Either would have made the top of her list. Instead of worrying about that, she zeroes in on the final relationship goal she had: convincing Sparks to be more open about his plans. She tries to recall the speech she was going to make, laying things out for Liz that she's convinced Sparks isn't sharing.

Before she can speak, though, Liz continues, saying, "I'm promoting you to VP of product and I'd like you to work with Sparks directly. Honestly," she adds, "based on what I'm hearing from all around the company, we should have hired you as a VP from the start."

While Irie processes this news, Liz keeps speaking, outlining upcoming events and what she needs from Irie. She wants Irie to help sell the company based on the value of what they've built and the insights they've gained about customers. Irie is about to interrupt when the full realization of what Liz has just said hits her.

"Wait," Irie says. "You knew Sparks was trying to sell the company?"

Liz stops speaking, confused, and looks at Irie. "Well, yes," she says. "I asked him to take the lead on this."

Shocked, Irie blurts out, "But he asked me not to tell you anything about it!"

Liz's eyes shift to Sparks and crinkle into a smile. "He tries a little too hard to protect me," she says.

"You need to focus on getting well," Sparks says. Irie hears him repeat this phrase he's said often over the last few months, but now hears it as genuine for the first time.

"You really have been trying to do what's best for Liz all this time," Irie says to Sparks. "I thought that was just an excuse for you to keep things to yourself."

"That was my doing, too," Liz says. "I asked Sparks to keep the extent of my condition quiet. I didn't want people to worry about me or about what my illness would mean for the future of the company."

Irie's mind is reeling as she rearranges the puzzle pieces of everything she's learned about Helthex, Sparks, and Liz into a new picture. Clearly she has more to learn about reading people, she thinks.

"I'm sorry I wasn't honest with you earlier," Liz continues. "Sparks convinced me that it was time to bring you in on things."

"I'm sorry I've been so difficult to work with," Sparks says. "I couldn't tell you much and that's hard for me. You were working a lot of it out anyway, I think."

"Clearly not everything," Irie says. "But why sell?" Irie continues, her concerns coming back clearly to mind. "We have something great going here."

"It's wonderful to hear you say that, Irie," Liz says, smiling. "We're going to need that passion to keep the company vision alive."

Irie says that is exactly what she's worried about. She explains everything Ella has uncovered about the plans these companies have for them and their product. She adds José's analysis about how much lower engagement is among B2B users. Liz and Sparks listen somberly, but Irie detects no surprise in their expressions. "You knew about these things, too," she says, disappointed.

Irie feels their vision of "personalized medical advice for anyone, anytime, anywhere" fading away as Liz says she thinks there is room for making progress toward that vision even within these companies. She asks Irie to stay strong and to use her influencing skills to do what she can.

Liz continues, but Irie is not really listening. She feels that Liz has given up on the vision but hasn't admitted it to herself. She sees her other reasons for staying fading, too, as people drift from the ghost of Helthex over time.

Irie's attention snaps back to the meeting when she hears Liz talking about how they would be able to keep the development team intact with either new parent. She realizes Liz hasn't really answered her question about why she wants to sell.

"We're running out of cash," Sparks interjects. "We're not going to get a better offer than the two we have now."

"I was going to go on the road and raise VC money," Liz says, "but I can't do that now, not soon enough. And I'm afraid public speaking isn't a strength of my friend here," she adds, smiling affectionately at Sparks. "No offense."

"None taken," says Sparks with a warding off gesture. "I rolled a 3 for charisma..."

Liz laughs, "...but an 18 for intelligence, thankfully."

"If we cut costs..." Irie begins, but Liz shakes her head.

"Philippe and I have been over and over the numbers. If we cut too much we can't keep moving forward and we fail. If we don't cut enough, we can't pay people and we fail. There is no path forward independently."

Irie wants to argue—to plead with Liz not to lose hope—but Liz's attention is drawn away by someone entering her hospital room. She nods to them and says she needs to sign off. The call ends and Irie finds herself facing Sparks alone.

She isn't sure what to make of her new interim CEO...and her new manager. He is a different person than she'd thought. His behavior has been awful, she reflects, but understanding the situation he was put into, she feels sympathy for him. She wonders how well she would have done in his shoes.

She feels once more that all of her quitting criteria have been met, but then she remembers she also gave herself six months to be sure. She also recalls Sri telling her he needs her help.

Sparks tells her to take some time to absorb all these changes. She tells him it is good advice, then silently decides not to take it. She calls a meeting of the product council for the following morning. *Maybe I do have a bit of Dominant in my decision style, too,* she thinks.

A week later, Irie is presenting an updated roadmap to Sparks. Liz is again dialing in, but she appears to be at home between treatments, properly dressed, and is using her laptop. Irie and the full product council are in a conference room with Sparks, except Liandri, who is remote as usual.

"We have developed a scenario we want to walk you through," Irie begins. "It's not the roadmap you asked for. I still have that if needed, but we—" she says with a gesture to the entire group "—wanted to provide an alternative."

In a few slides, Irie outlines a future version of the Helthex consumer app that is focused on actionable, in-the-moment, personalized advice for users with no overhead of administration or other business-focused capabilities. "José's analysis shows that we've reached product-market fit with these types of users, so we want to be the best solution for them, bar none." Liz is quiet, listening, so Irie continues.

"We looked into what this would take to build," she says "and to be honest, neither of these companies offers us a huge leg up in reaching this vision. Alvex's chatbot is very capable, but the hard part isn't the interactive chat capability. Yacob's team already has that working. The hard part is training the AI to provide useful, reliable, safe feedback to our users," she says, emphasizing the word "safe." "That's going to take time. Divya estimates six to nine months regardless."

Sparks frowns at the direction of the conversation. He asks if the data from VigGuard's subscribers would help with that problem.

"In the long run, a bigger training set would help optimize recommendations," Divya answers, "but probably only by ten to twenty percent. We have enough data from current users to train the model quite well."

"That's fine," interjects Sparks. "We can debate exactly how much each of these companies would or wouldn't help us, but we need a safe harbor."

Philippe speaks up then, saying he's worked with the team on some new cost projections. "The narrow focus of this plan allows us to be more strategic with our cuts," he explains. Irie puts up the budget slide.

Sergey explains he has developed a plan to slash the marketing spend until they're ready to launch and then do an invitation-only preview release. He believes the exclusivity would generate organic interest. Ella then adds that with this plan they could cut their enterprise sales team since they are currently unprofitable.

Philippe explains that lower acquisition rates in the meantime would also mean lower overall operational costs than projected.

Liandri says that she can postpone planned hiring in support, and she thinks her people will agree to a temporary hours reduction so she doesn't have to cut staff.

Sri says that he would repurpose some headcount in product, design, and engineering to Data Science to accelerate the timeline.

Liz speaks up for the first time. "This is amazing," she says. "How long could we last with this plan?"

"Unfortunately," Philippe says, "these cuts are still not enough."

"We could get something out—probably—by the end of the year, but we'd have nothing left," Irie says.

Liz looks to Philippe, and he confirms Irie's analysis.

"I'd like to look over the projections," says Sparks, looking directly at Irie, "but you have clearly done your homework."

"It wasn't my homework," Irie says. "It was the whole team."

Liz asks what it would take to fully commercialize this AI-powered version of the product. Irie gestures toward Philippe, who explains that they've run a number of Monte Carlo simulations based on assumptions from marketing. "If it goes well, and if we manage costs, we will be on a path to break even by the end of next year," he says. "We think we'd need another $7.5 million," he adds.

Liz looks disappointed, and Sparks throws up his hands, exasperated. "So we're right back where we started."

"We did a little digging on that, too," Ella says. "My contacts tell me that in addition to doing acquisitions, both companies have venture funds."

Phil then adds, "Four million from each would give us some extra margin and, at the valuation we've been discussing with them, we'd retain majority control."

Everyone is quiet then, looking at Sparks, who appears to be thinking hard. Seeming to relax, he then looks toward Liz on the wall monitor. "You've got enough to deal with," he says.

Liz has been listening with a growing sense of amazement. "We should have looked at this closer," she says. "This could really work."

Sparks is concerned. "I can't ask you to take on two fights at once," he says.

"What we've just heard convinces me that you and these people can handle this fight while I focus on my own," Liz replies. "And by the time we get to some level of traction, I could be ready to talk to investors again."

Sparks turns to Irie. "Do you think you could present this plan to the board?"

Irie considers the implied question in his request. Will she stay and work with him as her new boss?

"I'd like that," she replies. ■

Takeaways

Sometimes, you leverage all your stakeholder management skills but still find some relationships challenging enough that it's tempting to give up. And sometimes things change enough at your organization that you need to rethink whether it is still a fit for you.

- Check yourself first to see if you might be the problem in your stakeholder relationship. It's possible there are biases clouding your judgment of their motives. If you are the only one having a problem with this stakeholder, you may be able to learn something from peers with good relationships with them.
- Set relationship goals for you and your stakeholder. Focus on positive goals rather than getting sucked into drama or revenge. Focus only on what you can control.
- Set quitting criteria that include dates and actions if the criteria are not met. Quitting doesn't mean that you hand in your resignation that day—it can mean simply starting to actively look for other opportunities.

Your superpower is the collaborative culture you develop

Epilogue

Three years later…

"I'm so glad you connected me with your cousin," the barista says to Irie as she sets up a pourover for her. "His coffee is 25% of my business now."

"That makes me happy," says Irie, anticipating the warm brew. "And you make it much better than I do at home."

"Or at the office," says a familiar voice from behind Irie. Turning, she sees Sparks standing there looking unabashed.

Irie laughs. "That's true!" she says.

Taking their Jamaican coffees to a table by a window, Irie and Sparks sit down as they do every Monday before work.

"Friday was fun," Irie says in an ironic tone.

"The all-hands played out just like you said it would," Sparks says, seriously. "When you announced we were moving into wearables, there was a lot of excitement. You got Philippe to talk about the economics, Ella talked about cross-sell opportunities between the devices and the apps, Alex and José talked about user reactions to our research prototypes. I still can't believe you got Sergey excited about talking to an AI on his watch."

"I think the rigor we put into the process leading up to today helped," Irie says. "Going from idea to business plan to launching this whole initiative required a bunch of workshops, a whole new DACI, objectives, decision criteria, roadmap, and on and on. You even previewed some of this with our hardware partners to make sure they wouldn't all pull out of our app business if we competed with them. We were really prepared."

"We were really *aligned*," Sparks says, emphasizing that last word. "It was partly about process, sure, but it was a lot about the collaborative culture you've developed here—and you working across the company behind the scenes, providing context and direction. That's your superpower."

"You challenged me, helped me get better," she says, not wanting to take all the credit. "And we couldn't have done any of it without Liz's support."

Sparks agrees. "I'm so glad she came back," he says. "I was glad to step in as CEO for a while, but I'm much happier in this biz dev role. I can still work the partnerships and do a bit of prototype development on the side with my little team. I've always liked R&D."

"I've missed having Sri around, since he left," Irie says, returning to their earlier discussion. "He would have beamed in that meeting about the new product line and gotten his team revved up to work on it."

"It's true," Sparks replies. "Yacob doesn't have Sri's charisma—maybe he rolled a 10," he jokes.

"You've been a big help to Yacob," Irie says. "Since you've opened up more about what you're doing, he's come to look up to you."

"I don't know how good a mentor I am," Sparks says, "but I think he's doing well keeping things on course since Sri left. And he's more comfortable with Divya and José reporting to you."

Irie asks if Sparks has heard from their former CTO recently. Sri left a few months ago to build an AI-driven app to help people with physical therapy, something he's been using for years. "He raised a seed round and he hired Divya's brother to build a team."

"I thought those two didn't get along?" Irie says, surprised.

"Sri credits you for teaching him how to handle politics, actually," explains Sparks. "He said the trick was to assume the other person means well and listen to them."

Irie smiles, then sobers. "Speaking of which, did you have any interaction with our hot-headed tech lead after I talked with him and Pippa on Friday?"

"Nate?" Sparks asks. "He's still not convinced about Pippa. What set the two of them off anyway? Did I miss something when I left the all-hands meeting early?"

Irie recounts what happened during the Q&A session at the end of Friday's all-hands meeting. Irie introduced Pippa as the new director of product for their consumer app. Pippa is replacing Christina, who rose to lead their flagship product but is now moving over to take on their new wearables product line. Nate asked how Pippa plans to handle security given all the news about quantum computers breaking conventional encryption.

Irie says that she wishes Pippa had either referred the question to Yacob, or better yet asked him about his concerns. "Instead, she just said she wasn't familiar with the problem but would look into it."

Sparks looks questioningly at her. "You would normally say that admitting what you don't know helps with credibility, wouldn't you?"

"Well, that didn't work with Nate," Irie says with a sigh. "She didn't know him, of course, but you know he looks down on anyone without a computer science degree. He came right out and said we should've hired someone more technical."

Sparks merely raises an eyebrow.

"And Pippa didn't take that well," Irie continues. "She was clearly embarrassed, she got defensive, and started talking about her past successes. I shut the conversation down at that point and asked them both to meet with me later."

She sighs and admits, "It was her first day, and I couldn't help thinking of my first day and my first meeting with you."

Sparks looks uncomfortable, but Irie continues. "I don't think I handled that meeting well, and I don't think Pippa did great here, either. So that's why I want you to help me speak to them. You and I butted heads a lot during my first few months, but we worked it out."

"I'm not proud of how difficult I made all of that for you," Sparks says.

"And I was too proud to put my ego aside and see that there was more I needed to know," Irie replies. "I was ready to quit when Liz told me you were the new CEO and I was supposed to report to you. Nate and Pippa are getting along just as badly as we were then."

Irie proposes that she coach Pippa and that Sparks offer to coach Nate in stakeholder management. Surprised, Sparks asks, "Why me? And what if he doesn't want a coach?" he adds.

Irie responds that he can make it clear how important teamwork is at Helthex, and that he can connect with Nate "difficult to difficult." This makes Sparks grin. "I'll do it," he says, "but only if you send me that secret Product Playbook of yours." Irie agrees.

"Actually, this fits with some plans I wanted to tell you about," Sparks says. He explains that he is going to leave the company after their planned IPO. He feels he's accomplished what he wanted to and he doesn't enjoy working in the expanded Helthex. "It'll take me six months to wind down," he adds. "That'll give me time to coach Nate."

"Also, Chloë is pregnant," he reveals, "so I'm going to want more time at home."

Irie congratulates him warmly and asks what he will do next. "I was thinking about advising, like your friend, Darius," he explains.

"Arianna can handle the partners," he says, "but I'd like her to report to you. And," he adds, "I'd like you to take my little R&D team."

Irie is intrigued but also a bit apprehensive. She's worked on partnerships through Christina. It's more to deal with but she thinks she can handle it. "That zero-to-one stuff is not my strong suit," she says. "I'm more of a scaler-upper. That's why that R&D team reports to you."

Sparks points out that she's worked closely with him and the team on the development of this wearables initiative. "Eitan works for you and the other people know you well. They enjoy working with you," he adds. "More than they do working with me, I think."

He points out that what they all need most is context and direction, her "superpower." He makes the case that the partnerships and "this zero-to-one stuff" are now just part of their growing portfolio of product investments. "You've already got, what, fifty people? This is just a few more."

Irie says managing all that sounds like a big job. "Yes," replies Sparks. "And you are the best person for it. That's why I think we should talk to Liz about making you CPO."

"I'm going to need a whole new playbook!" Irie says. ■

Index

D

E

F

G

Q

R

S

T

U

V

W

About the Authors

Bruce McCarthy, founder of Product Culture, helps companies like NewStore, Camunda, hyperexponential, Socure, and Toast achieve their product visions through advising, coaching, and workshops.

Melissa Appel coaches product management leaders, helping them build and manage effective teams and improve stakeholder relationships. She previously spent 20 years as a practitioner at companies of various sizes and stages.

Acknowledgments

First of all, Bruce and Melissa would like to thank our book designer, Michael Connors, without whom this book would not be easy to look at, navigate, or understand. (No one wants to see Bruce and Melissa's initial drafts of the figures.)

We'd also like to thank the folks at O'Reilly who helped us bring the whole thing together, including but not limited to Amanda, Angela, Kristen, Suzanne, Ron, and Susan, plus our freelance editors, Liz and James, and our indexer, Joanne. We couldn't have done it without you.

We would like to make special mention of the invaluable contributions to this book from our Early Readers Club. Over 140 amazing people loaned us their time, ears, ideas, anecdotes, quotes, and endless rounds of feedback and encouragement over the two years it took to write (and rewrite) this book. The first draft of this book wasn't all we wanted it to be. You gave us the tough love we needed and helped us make it better. We thank you.

We'd like to especially call out:

- Jerry Odenwelder for openly sharing his experiences, tips, and challenges
- Keith Hopper for challenging us to show the positives in each stakeholder
- Ashlyn Baum for insisting we name the different sorts of stakeholders
- Christina Wodtke for inspiring us toward diversity in the characters

- Daniele Beccari for validating some of our early thinking about the Helthex app
- Gautier Scherrer for wanting to hear the character's accents
- Helen Saunders for anecdotes of dysfunctional organizations
- Joshua Herzig-Marx for always asking good questions and connecting us with Lenny
- Kevin Segedi for sharing his new job, new company experience month by month
- Liz Lehtonen for vividly describing the introvert's experience (and tips)
- Meaghan Swain for being one of our first book club members
- Nick Katis for reminding us how personal all of this work is
- Tim Bohour for sharing some the scariest things he's seen at work
- Trace Wax for his candor with a smile
- Warren Brown for his candor and sardonic humor
- C. Todd Lombardo for being there for us all along
- Tim Bouhour for trusting us with his personal experiences
- Matt Kaplan for articulating the world of uncertainty that is product
- Sheyda Esmaeilnejad for her comprehensive feedback
- Shawn Myers for sharing his personal journey with stakeholders
- John Toklu for sharing his biggest challenges

Thank you to our amazing book reviewers, who got through the whole book and gave us amazing feedback: Adam Thomas, Afonso Malo Franco, EJ Ogenyi, Gabriell Washington, Gautier Scherrer, Helen Saunders, Jerry Odenwelder, Joshua Herzig-Marx, Keith Hopper, Matt Price, Meaghan Swain, Michael Pierce, Sandy Guillot, and Tim Bouhour. And especially C. Todd Lombardo, Liz Lehtonen, and Sheyda Esmaeilnejad who each read through both draft versions.

Thank you to the rest of our Early Readers Club members: Adrian Howard, Alexander Harwitz, Alice Mitchell, Amanda Shipka, Anabela Cesário, Andrei Maxwel Mulbauer, Andreia Soares, André Vieira, Ângela Dinis, Angela Ilas, Anne Stuneck, Annie Hooper, Benjamin Coughlin, Brendan Wightman, Brian Levenson, Bruno Fernandes, Caleb Yell, Casey Karye-Haddy, Daerson Oliveira, Dan Iorg, Dan Wolman, Daniel Araujo, Daniel Mitchell,

David Ehringer, Devon McCarthy, Dilip Hari, Emma Lopez, Erin Teare, Felipe Castro, Fernando Moitinho, Filipe Rocha, Florian Zilliox, Frederico Pires, Helio Cardoso, Hugo Gonçalves, Iryna Melnyk, Jen Leibhart, Joe Marini, Jonathan Abbett, Jonathan Coughlin, Joseph Spooner, João Bento, João Romão, João Santos, Julia Steier, Kathy Martins, Kevin Doole, Lambert Bouley, Lara Leite, Lars Bönke, Lavanya Vipin, Lisa Insley, Lynne Levy, Madeline Wang, Madhu Vulpala, Manuel Pereira, Marc Abraham, Marc-Andre Ferguson, Mardien Drew, Mark Bussell, Martin Slaney, Martina Hodges-Schell, María Barriocanal, Matt Price, Melissa Johnson, Michael Hyde, Nuno Pereira, Paulo Alves, Phil Garner, Pradeep Thathachari, Randy Smith, Ray Cha, Ricardo Alves, Sam Gledhill, Sara Santos, Scott Piccolo, Sean McAuliffe, Sharmila Gopirajan, Sharon Duan, Solange Ferreira, Sree Kamireddy, Stephen Castro-Starkey, Stephen Sulik, Sue Hogg, Tatiana Leon, Tiago Leão, Tom Booth, Tom Kazer, Tomas Halgas, Tyler Barnes, Varun Maheshwari, Vasco Pessanha, Vivek Chandrashekar, and Wes Bush.

We would also like to send eternal gratitude to Phil Hornby for being an unofficial editor and contributor to our many drafts and stepping in for us when we were buried in revisions, to Nina Mitchell for getting the word out and orchestrating everything with a smile and a laugh, to Win Rosales for making everything easier every day, and to Frank, Nuno, Ed, and Alfonso for your relentless support.

Bruce would also like to thank his very understanding wife, Christine, his unerring copyeditor daughter, Amanda, and his empathetic daughter, Devon, with the amazing ear. He would also like to thank his coauthor, Melissa, for proposing they write a book together and for always pushing to make it better. He'd like to thank Medusa Brewing and Acton Coffee House also for taking such good care of him while he struggled to get the wording just right.

Melissa would like to thank her husband, JP Nahmias, and her girls, Rose, Maya, and Lily, who put up with her working on the book on too many evenings and weekends. And she'd like to thank Bruce for always challenging her and for being an expert novelist.

Milton Keynes UK
Ingram Content Group UK Ltd.
UKHW050102280824
447505UK00001B/2